T0339126

HARVARD-YENCHING INSTITUTE STUDIES
XXI

STUDIES IN CHINESE LITERATURE

Edited by
JOHN L. BISHOP

HARVARD UNIVERSITY PRESS
Cambridge, Massachusetts
London, England

ISBN 0-674-84705-9

Library of Congress Catalog Card Number 65-13836

PRINTED IN THE UNITED STATES OF AMERICA

FOREWORD

This volume, the twenty-first of the Harvard-Yenching Institute Studies, is financed from the residue of the funds granted during World War II by the Rockefeller Foundation for the publication of Chinese and Japanese dictionaries. This series is distinct from the Harvard-Yenching Institute Monograph Series and consists primarily of bibliographical studies, grammars, reference works, translations, and other study and research aids.

PREFACE

While specialized studies in Chinese literature multiply, an adequate history of Chinese literature based upon such studies has still to be written. It is, therefore, necessary to keep those preliminary materials accessible. Most of the eight articles reprinted in this volume have been unavailable for some time, and their reissue has been undertaken to fulfill a demand which is, if not vociferous, at least persistent. Six of the studies first appeared in issues of the *Harvard Journal of Asiatic Studies* that are now out of print. The editor is grateful to Ejnar Munksgaard, Ltd, publisher of *Studia Serica Bernhard Karlgren Dedicata*, for permission to include James R. Hightower's "Some Characteristics of Parallel Prose" and to *The Journal of Asian Studies* for permission to reprint his own "Some Limitations of Chinese Fiction" from *The Far Eastern Quarterly*.

In the present publication, original page numbers have been retained at the top of the page and continuous page numbers have been added at the foot of the page. Additions and corrections (in which reference is made to the original pagination) have been listed at the end of each article and a list of abbreviations used in the text may be found facing page 1.

J. L. B.

November 1, 1964
Cambridge, Massachusetts

CONTENTS

ABBREVIATIONS

BSOAS	*Bulletin of the School of Oriental and African Studies*
HJAS	*Harvard Journal of Asiatic Studies*
JAOS	*Journal of the American Oriental Society*
MN	*Monumenta Nipponica*
SBE	*Sacred Books of the East* Series
SPPY	*Ssu-pu pei-yao*
SPTK	*Ssu-pu ts'ung-k'an*
TP	*T'oung pao*
TSCC	*Ts'ung-shu chi-ch'eng*

RHYMEPROSE ON LITERATURE

THE *WÊN-FU* OF LU CHI (A.D. 261–303)

Translated and Annotated by
ACHILLES FANG

RHYMEPROSE ON LITERATURE
THE *WÊN-FU* OF LU CHI (A. D. 261–303)
陸機:文賦

TRANSLATED AND ANNOTATED

BY

ACHILLES FANG

———

CONTENTS

INTRODUCTION

TU FU thinks LU Chi wrote the *Wên-fu* when he was twenty years old: 陸機二十作文賦 (cf. 醉歌行 in *Collected Poems, Ssŭ-pu ts'ung-k'an* ed. 9.16a). Ho Ch'o 何焯 (義門讀書記, 一文選), however, writes that the poet misinterpreted TSANG Jung-hsü's 臧榮緒 statement quoted in the *Wên-hsüan* commentary of LI Shan 李善; in his book purporting to correct Ho Ch'o's errors, Hsü P'an-fêng 徐攀鳳(選學糾何) defends TU Fu. All we can say, then, is that LU Chi wrote down the 1658 characters of his rhymeprose * on literature sometime before he was killed in A. D. 303, *aetat.* 43.

This compact essay is considered one of the most articulate treatises on Chinese poetics. The extent of its influence in Chinese literary history is equalled only by that of the more comprehensive sixth-century work, *Wên-hsin tiao-lung* 文心雕龍 of LIU Hsieh 劉勰.

———

* "Rhymeprose" is derived from "Reimprosa" of German medievalists.

3

The *Wên-fu* proper consists of 131 distichs, mostly parallel
lines or antithetical couplets. A large majority of these distichs,
105 to be exact, are in six-character lines; couplets in four-
character lines amount to 17 only (§§ 8, 9, 17, 36, 37, 63, 67, 68,
72, 73, 94, 95, 111, 112, 113, 118, 119). Of the remaining 9
distichs, 5 are in five-character lines (§§ 18, 41, 42, 59, 62); one
each in seven- (§ 99), eight- (§ 100), and fifteen-character lines
(§ 13); and finally there is one irregular couplet (§ 124), which
may be classed as a variation of the eight-character line. The
term " Four-and-Six Prose " (*ssŭ-liu wên* 四六文), then, applies
to the *Wên-fu*, for the two classes combined leave only 9 distichs
out of 131 unaccounted for; although four-character lines are not
so numerous as the six-character ones.

As the preface in normal prose comprises 120 characters, while
the distichs amount to 1513 characters and the transition words
(§§ 8, 10, 13, 17, 41, 67, 72, 94, 99, 111, 118, 122) to 25, the
text of the entire *Wên-fu* as I have established it comprises 1658
characters.

This is not the first translation of the *Wên-fu*: Dr. G. MAR-
GOULIÈS's first French version appeared in 1926 (*Le " Fou " dans
le Wen-siuan*, Paris, 82-97) and his second in 1948 (*Anthologie
raisonnée de la littérature chinoise*, Paris, 419-425); the late B. M.
ALEXÉIEV published his Russian rendition in 1944 (*Bulletin de
l'Académie des Sciences de l'URSS*, " Classe des sciences littéraires
et linguistiques," 3 (4).143-64); Mr. CH'ÊN Shih-hsiang and Mr.
E. R. HUGHES have each given us an English translation, the
former in his study entitled " Literature as Light against Dark-
ness " (National Peking University Semicentennial Papers No.
11, College of Arts, Peiping, 1948) and the latter in mimeographed
form, privately circulated.* Accepting Mr. Bernard BERENSON's
challenge, " Then dare to translate the ancient Chinese and
Indian thinkers " (*Sketch for a Self-Portrait*), I felt that I had
to make my translation independently on the basis of " my little

---

* [EDITOR'S NOTE: After this manuscript was received from the author, Mr. HUGHES'
The Art of Letters, Lu Chi's " Wen Fu," A. D. 302, was published in the Bollingen
Series. See Mr. FANG's review on pages 615-636 of this issue of *HJAS*.]

psychosinology." Hence, I have not made use of the previous translations, excellent as they seem to be.

The subtitles are my own contribution. That the subdivision is not arbitrary can be seen from the rhyme scheme (see Appendix I). For the sake of the reader who might be puzzled at the apparently inconsistent use of pronouns in my translation, it may be here stated that they are all interpolated by me, except in

> Preface § 1:   " I " (余)
>   Text § 70: " my heart " (余懷)
>   " me " (我)
>   "   § 101: " I " (余)
>   "   § 106: " my two hands " (予掬)
>   "   § 123: " my power " (余力)
>   "   § 124: " I " (吾).

I employ " you " and " he " and even " we " indiscriminately in accordance with my sense of rhythm.

I am grateful to Mr. Archibald MacLeish, the Boylston Professor at Harvard University, for the interest he has taken in my translation. If it is at all readable, it is due in great measure to Mr. MacLeish.

As far as notes go, I am at one with a contemporary of Rousseau's: " Il faut dire en deux mots / Ce qu'on veut dire; / Les long propos / sont sots.∥ Il ne faut pas toujours conter, / Citer / Dater / Mais écouter. . . ." But I cannot claim " J'ai réussi," especially because I broke Mme. de Boufflers' injunction (" Il faut éviter l'emploi / Du moi, du moi "). At any rate, modesty commands me to relegate my notes to the limbo of appendices.

Translation

Rhymeprose on Literature

Preface

(§§ 1-10)

(in unrhymed prose)

余每觀才士之所作竊有以得其用心

Each time I study the works of great writers, I flatter myself
I know how their minds worked.

夫其放言遣辭良多變矣

Certainly expression in language and the charging of words with
meaning can be done in various ways.

妍蚩好惡可得而言

Nevertheless we may speak of beauty and ugliness, of good and
bad [in each literary work].

每自屬文尤見其情

Whenever I write myself, I obtain greater and greater insight.

恒患意不稱物文不逮意

Our constant worry is that our ideas may not equal their objects
and our style may fall short of our ideas.

蓋非知之難能之難也

The difficulty, then, lies not so much in knowing as in doing.

故作文賦以述先士之盛藻因論作文之利害所由

I have written this rhymeprose on literature to expatiate on the
consummate artistry of writers of the past and to set forth
the whence and why of good and bad writings as well.

他日殆可謂曲盡其妙

May it be considered, someday, an exhaustive treatment.

至於操斧伐柯雖取則不遠者夫隨手之變良難以辭逮

Now, it is true, I am hewing an ax handle with an ax handle in my
hand: the pattern is not far to seek.

6

However, the conjuring hand of the artist being what it is, I
    cannot possibly make my words do the trick.

蓋所能言者具於此云

Nevertheless, what I am able to say I have put down here.

Text

A

(§§ 1-7)

Preparation

佇中區以玄覽　頤情志於典墳

Taking his position at the hub of things, [the writer] contem-
    plates the mystery of the universe; he feeds his emotions
    and his mind on the great works of the past.

遵四時以歎逝　瞻萬物而思紛

Moving along with the four seasons, he sighs at the passing of
    time; gazing at the myriad objects, he thinks of the com-
    plexity of the world.

悲落葉於勁秋　喜柔條於芳春

He sorrows over the falling leaves in virile autumn; he takes joy
    in the delicate bud of fragrant spring.

心懍懍以懷霜　志眇眇而臨雲

With awe at heart, he experiences chill; his spirit solemn, he turns
    his gaze to the clouds.

詠世德之駿烈　誦先民之清芬

He declaims the superb works of his predecessors; he croons the
    clean fragrance of past worthies.

游文章之林府　嘉麗藻之彬彬

He roams in the Forest of Literature, and praises the symmetry of
    great art.

慨投篇而援筆　聊宣之乎斯文

Moved, he pushes his books away and takes the writing-brush,
    that he may express himself in letters.

7

B

(§§ 8-16)

PROCESS

其始也　皆　收視反聽　耽思傍訊

At first he withholds his sight and turns his hearing inward; he is
　　lost in thought, questioning everywhere.

精騖八極　心游萬仞

His spirit gallops to the eight ends of the universe; his mind
　　wanders along vast distances.

其致也　情瞳曨而彌鮮　物昭晰而互進

In the end, as his mood dawns clearer and clearer, objects, clean-
　　cut now in outline, shove one another forward.

傾羣言之瀝液　漱六藝之芳潤

He sips the essence of letters; he rinses his mouth with the extract
　　of the Six Arts.

浮天淵以安流　濯下泉而潛浸

Floating on the heavenly lake, he swims along; plunging into the
　　nether spring, he immerses himself.

於是　沈辭怫悅若游魚銜鈎而出重淵之深　浮藻聯翩若翰鳥纓繳而
墮曾雲之峻

Thereupon, submerged words wriggle up, as when a darting fish,
　　with the hook in its gills, leaps from a deep lake; floating
　　beauties flutter down, as when a high-flying bird, with
　　the harpoon-string around its wings, drops from a crest
　　of cloud.

收百世之闕文　採千載之遺韻

He gathers words never used in a hundred generations; he picks
　　rhythms never sung in a thousand years.

謝朝華於已披　啓夕秀於未振

He spurns the morning blossom, now full blown; he plucks the
　　evening bud, which has yet to open.

觀古今於須臾　撫四海於一瞬

He sees past and present in a moment; he touches the four seas
　　in the twinkling of an eye.

8

C

(§§ 17-29)

Words, Words, Words

然後 選義按部 考辭就班
Now he selects ideas and fixes them in their order; he examines
words and puts them in their places.

抱景者咸叩 懷響者畢彈
He taps at the door of all that is colorful; he chooses from among
everything that rings.

或因枝以振葉 或沿波而討源
Now he shakes the foliage by tugging the twig; now he follows
back along the waves to the fountainhead of the stream.

或本隱以之顯 或求易而得難
Sometimes he brings out what was hidden; sometimes, looking for
an easy prey, he bags a hard one.

或虎變而獸擾 或龍見而鳥瀾
Now, the tiger puts on new stripes, to the consternation of other
beasts; now, the dragon emerges, and terrifies all the birds.

或妥帖而易施 或岨峿而不安
Sometimes things fit together, are easy to manage; sometimes they
jar each other, are awkward to manipulate.

罄澄心以凝思 眇衆慮而爲言
He empties his mind completely, to concentrate his thoughts; he
collects his wits before he puts words together.

籠天地於形內 挫萬物於筆端
He traps heaven and earth in the cage of form; he crushes the
myriad objects against the tip of his brush.

始躑躅於燥吻 終流離於濡翰
At first they hesitate upon his parched lips; finally they flow
through the well-moistened brush.

9

理扶質以立幹　文垂條而結繁

Reason, supporting the matter [of the poem], stiffens the trunk; style, depending from it, spreads luxuriance around.

信情貌之不差　故每變而在顏

Emotion and expression never disagree: all changes [in his mood] are betrayed on his face.

思涉樂其必笑　方言哀而已歎

If the thought touches on joy, a smile is inevitable; no sooner is sorrow spoken of than a sigh escapes.

或操觚以率爾　或含毫而邈然

Sometimes words flow easily as soon as he grasps the brush; sometimes he sits vacantly, nibbling at it.

D

(§§ 30-35)

Virtue

伊茲事之可樂　固聖賢之所欽

There is joy in this vocation; all sages esteem it.

課虛無以責有　叩寂寞而求音

We [poets] struggle with Non-being to force it to yield Being; we knock upon Silence for an answering Music.

函緜邈於尺素　吐滂沛乎寸心

We enclose boundless space in a square foot of paper; we pour out a deluge from the inch-space of the heart.

言恢之而彌廣　思按之而逾深

Language spreads wider and wider; thought probes deeper and deeper.

播芳蕤之馥馥　發青條之森森

The fragrance of delicious flowers is diffused; exuberant profusion of green twigs is budding.

粲風飛而猋豎　鬱雲起乎翰林

A laughing wind will fly and whirl upward; dense clouds will arise from the Forest of Writing Brushes.

10

E

(§§ 36-49)

DIVERSITY

(i) *The Poet's Aim*

(§§ 36-42)

體有萬殊　物無一量

Forms vary in a thousand ways; objects are not of one measure.

紛紜揮霍　形難爲狀

Topsy-turvy and fleeting, shapes are hard to delineate.

辭程才以效伎　意司契而爲匠

Words vie with words for display, but it is mind that controls them.

在有無而僶俛　當淺深而不讓

Confronted with bringing something into being or leaving it unsaid, he groans; between the shallow and the deep he makes his choice resolutely.

雖離方而遯員　期窮形而盡相

He may depart from the square and deviate from the compasses; for he is bent on exploring the shape and exhausting the reality.

故夫　誇目者尙奢　愜心者貴當

Hence, he who would dazzle the eyes makes much of the gorgeous; he who intends to convince the mind values cogency.

言窮者無隘　論達者唯曠

If persuasion is your aim, do not be a stickler for details; when your discourse is lofty, you may be free and easy in your language.

(ii) *Genres*

(§§ 43-49)

詩緣情而綺靡　賦體物而瀏亮
Shih (lyric poetry) traces emotions daintily; *Fu* (rhymeprose) embodies objects brightly.

碑披文以相質　誄纏緜而悽愴
Pei (epitaph) balances substance with style; *Lei* (dirge) is tense and mournful.

銘博約而溫潤　箴頓挫而清壯
Ming (inscription) is comprehensive and concise, gentle and generous; *Chên* (admonition), which praises and blames, is clear-cut and vigorous.

頌優游以彬蔚　論精微而朗暢
Sung (eulogy) is free and easy, rich and lush; *Lun* (disquisition) is rarified and subtle, bright and smooth.

奏平徹以閑雅　說煒曄而譎誑
Tsou (memorial to the throne) is quiet and penetrating, genteel and decorous; *Shuo* (discourse) is dazzling bright and extravagantly bizarre.

雖區分之在茲　亦禁邪而制放
Different as these forms are, they all forbid deviation from the straight, and interdict unbridled license.

要辭達而理舉　故無取乎冗長
Essentially, words must communicate, and reason must dominate; prolixty and long-windedness are not commendable.

F

(§§ 50-56)

Multiple Aspects

其爲物也多姿　其爲體也屢遷
As an object, literature puts on numerous shapes; as a form, it undergoes diverse changes.

12

其會意也尙巧　其遣言也貴妍

Ideas should be cleverly brought together; language should be
　　beautifully commissioned.

暨音聲之迭代　若五色之相宣

And the mutation of sounds and tones should be like the five
　　colors of embroidery sustaining each other.

雖逝止之無常　固崎錡而難便

It is true that your moods, which come and go without notice,
　　embarrass you by their fickleness,

苟達變而識次　猶開流以納泉

But if you can rise to all emergencies and know the correct order,
　　it will be like opening a channel from a spring of water.

如失機而後會　恒操末以續顚

If, however, you have missed the chance and reach the sense be-
　　latedly, you will be putting the tail at the head.

謬玄黃之秩序　故淟涊而不鮮

The sequence of dark and yellow being deranged, the whole
　　broidery will look smudged and blurred.

G

REVISION

(§§ 57-61)

或仰逼於先條　或俯侵於後章

Now you glance back and are constrained by an earlier passage;
　　now you look forward and are coerced by some anticipated
　　line.

或辭害而理比　或言順而義妨

Sometimes your words jar though your reasoning is sound, some-
　　times your language is smooth while your ideas make
　　trouble;

離之則雙美　合之則兩傷

Such collisions avoided, neither suffers; forced together, both
　　suffer.

13

考殿最於錙銖　定去留於毫芒

Weight merit or demerit by the milligram; decide rejection or
retention by a hairbreadth.

苟銓衡之所裁　固應繩其必當

If your idea or word has not the correct weight, it has to go,
however comely it may look.

H

Key Passages

(§§ 62-66)

或文繁理富　而意不指適

Maybe your language is already ample and your reasoning rich,
yet your ideas do not round out.

極無兩致　盡不可益

If what must go on cannot be ended, what has been said in full
cannot be added to.

立片言而居要　乃一篇之警策

Put down terse phrases here and there at key positions; they
will invigorate the entire piece.

雖衆辭之有條　必得兹而效績

Your words will acquire their proper values in the light of these
phrases.

亮功多而累寡　故取足而不易

This clever trick will spare you the pain of deleting and excising.

I

(§§ 67-71)

Plagiarism

或　藻思綺合　清麗千眠

It may be that language and thought blend into damascened
gauze—fresh, gay, and exuberantly lush;

14

炳若縟繡　悽若繁絃

Glowing like many-colored broidery, mournful as multiple chords;

必所擬之不殊　乃闇合乎曩篇

But assuredly there is nothing novel in my writing, if it coincides with earlier masterpieces.

雖杼軸於予懷　怵他人之我先

True, the arrow struck my heart; what a pity, then, that others were struck before me.

苟傷廉而愆義　亦雖愛而必捐

As plagiarism will impair my integrity and damage my probity, I must renounce the piece, however fond I am of it.

J

(§§ 72-78)

PURPLE PATCHES

或　若發穎豎　離衆絕致

It may be that one ear of the stalk buds, its tip standing prominent, solitary and exquisite.

形不可逐　響難爲係

But shadows cannot be caught; echoes are hard to bind.

塊孤立而特峙　非常音之所緯

Standing forlorn, your purple passage juts out conspicuously; it can't be woven into ordinary music.

心牢落而無偶　意徘徊而不掃

Your mind, out of step, finds no mate for it; your ideas, wandering hither and thither, refuse to throw away that solitary passage.

石韞玉而山暉　水懷珠而川媚

When the rock embeds jade, the mountain glows; when the stream is impregnated with pearls, the river becomes alluring.

彼榛楛之勿翦　亦蒙榮於集翠

When the hazel and arrow-thorn bush is spared from the sickle, it will glory in its foliage.

15

綴下里於白雪　吾亦濟夫所偉

We will weave the market ditty into the classical melody; perhaps
we may thus rescue what is beautiful.

K

(§§ 79-93)

FIVE IMPERFECTIONS

(i) *In Vacuo*

(§§ 79-81)

或託言於短韻　對窮迹而孤興

Maybe you have entrusted your diction to an anemic rhythm;
living in a desert, you have only yourself to talk to.

俯寂寞而無友　仰寥廓而莫承

When you look down into Silence, you see no friend; when you
lift your gaze to Space, you hear no echo.

譬偏絃之獨張　含清唱而靡應

It is like striking a single chord—it rings out, but there is no music.

(ii) *Discord*

(§§ 82-84)

或寄辭於瘁音　言徒靡而弗華

Maybe you fit your words to a frazzled music; merely gaudy, your
language lacks charm.

混姸蚩而成體　累良質而爲瑕

As beauty and ugliness are commingled, your good stuff suffers.

象下管之偏疾　故雖應而不和

It is like the harsh note of a wind instrument below in the courtyard; there is music, but no harmony.

16

(iii) *Novelty for Novelty's Sake*

(§§ 85-87)

或遺理以存異　徒尋虛以逐微

Maybe you forsake reason and strive for the bizarre; you are
　　merely searching for inanity and pursuing the trivial.

言寡情而鮮愛　辭浮漂而不歸

Your language lacks sincerity and is poor in love; your words
　　wash back and forth and never come to the point.

猶絃幺而徽急　故雖和而不悲

They are like a thin chord violently twanging—there is harmony,
　　but it is not sad.

(iv) *License*

(§§ 88-90)

或奔放以諧合　務嘈囋而妖冶

Maybe by galloping unbridled, you make your writing sound well;
　　by using luscious tunes, you make it alluring.

徒悅目而偶俗　固聲高而曲下

Merely pleasing to the eye, it mates with vulgarity—a fine voice,
　　but a nondescript song.

寤防露與桑間　又雖悲而不雅

It reminds one of Fang-lu and Sang-chien,—it is sad, but not
　　decorous.

(v) *Insipidity*

(§§ 91-93)

或清虛以婉約　每除煩而去濫

Or perhaps your writing is simple and terse, all superfluities
　　removed—

闕大羹之遺味　同朱絃之清氾

So much so that it lacks even the lingering flavor of a sacrificial
　　broth; it rather resembles the limpid tune of the " ver-
　　milion chord."

17

雖一唱而三歎　固旣雅而不豔

" One man sings, and three men do the refrain "; it is decorous, but it lacks beauty.

L

(§§ 94-100)

VARIABILITY

若夫　豐約之裁　俯仰之形

As to whether your work should be loose or constricted, whether you should mould it by gazing down or looking up,

因宜適變　曲有微情

You will accommodate necessary variation, if you would bring out all the overtones.

或言拙而喻巧　或理朴而辭輕

Maybe your language is simple, whereas your conceits are clever; maybe your reasoning is plain, but your words fall too lightly.

或襲故而彌新　或沿濁而更清

Maybe you follow the beaten track to attain greater novelty; maybe you immerse yourself in the muddy water—to reach true limpidity.

或覽之而必察　或研之而後精

Well, perspicacity may come after closer inspection; subtlety may ensue from more polishing.

譬猶　舞者赴節以投袂　歌者應絃而遺聲

It is like dancers flinging their sleeves in harmony with the beat or singers throwing their voices in tune with the chord.

是蓋輪扁所不得言　亦非華說之所能明

All this is what the wheelwright P'ien despaired of ever explaining; it certainly is not what mere language can describe.

18

M

(§§ 101-106)

MASTERPIECES

普辭條與文律　良余膺之所服

I have been paying tribute to laws of words and rules of style.

練世情之常尤　識前脩之所淑

I know well what the world blames, and I am familiar with what the worthies of the past praised.

雖濬發於巧心　或受蚩於拙目

Originality is a thing often looked at askance by the fixed eye.

彼瓊敷與玉藻　若中原之有菽

The fu-gems and jade beads, they say, are as numerous as the " pulse in the middle of the field " [which everyone can pick].

同橐籥之罔窮　與天地乎並育

As inexhaustible as the space between heaven and earth, and growing co-eternally with heaven and earth themselves.

雖紛藹於此世　嗟不盈於予掬

The world abounds with masterpieces; and yet they do not fill my two hands.

N

(§§ 107-110)

THE POET'S DESPAIR

患挈瓶之屢空　病昌言之難屬

How I grieve that the bottle is often empty; how I sorrow that Elevating Discourse is hard to continue.

故踸踔於短韻　放庸音以足曲

No wonder I limp along with trivial rhythms and make indifferent music to complete the song.

19

恒遺恨以終篇　豈懷盈而自足

I always conclude a piece with a lingering regret; can I be smug and self-satisfied?

懼蒙塵於叩缶　顧取笑乎鳴玉

I fear to be a drummer on an earthen jug; the jinglers of jade pendants will laugh at me.

O

(§§ 111-124)

INSPIRATION

(i)

若夫　應感之會　通塞之紀

As for the interaction of stimulus and response, and the principle of the flowing and ebbing of inspiration,

來不可遏　去不可止

You cannot hinder its coming or stop its going.

藏若景滅　行猶響起

It vanishes like a shadow, and it comes like echoes.

方天機之駿利　夫何紛而不理

When the Heavenly Arrow is at its fleetest and sharpest, what confusion is there that cannot be brought to order?

思風發於胸臆　言泉流於唇齒

The wind of thought bursts from the heart; the stream of words rushes through the lips and teeth.

紛葳蕤以馺遝　唯毫素之所擬

Luxuriance and magnificence wait the command of the brush and the paper.

文徽徽以溢目　音泠泠而盈耳

Shining and glittering, language fills your eyes; abundant and overflowing, music drowns your ears.

20

(ii)

及其　六情底滯　志往神留

When, on the other hand, the Six Emotions become sluggish and foul, the mood gone but the psyche remaining,

兀若枯木　豁若涸流

You will be as forlorn as a dead stump, as empty as the bed of a dry river.

攬營魂以探賾　頓精爽於自求

You probe into the hidden depth of your soul; you rouse your spirit to search for yourself.

理翳翳而逾伏　思乙乙其若抽

But your reason, darkened, is crouching lower and lower; your thought must be dragged out by force, wriggling and struggling.

是以　或竭情而多悔　或率意而寡尤

So it is that when your emotions are exhausted you produce many faults; when your ideas run freely you commit fewer mistakes.

雖茲物之在我　非余力之所勠

True, the thing lies in me, but it is not in my power to force it out.

故時撫空懷而自惋　吾未識夫開塞之所由

And so, time and again, I beat my empty breast and groan; I really do not know the causes of the flowing and the not flowing.

P

(§§ 125-131)

CODA: ENCOMIUM

伊茲文之爲用　固衆理之所因

The function of style is, to be sure, to serve as a prop for your ideas.

(*Yet allow me to expatiate on the art of letters:*)

21

恢萬里而無閡　通億載而爲津

It travels over endless miles, removing all obstructions on the way;
　　　it spans innumerable years, taking the place, really, of a
　　　bridge.

俯貽則於來葉　仰觀象乎古人

Looking down, it bequeaths patterns to the future; gazing up, it
　　　contemplates the examples of the ancients.

濟文武於將墜　宣風聲於不泯

It preserves the way of Wên and Wu, about to fall to the ground;
　　　and it propagates good ethos, never to perish.

塗無遠而不彌　理無微而不綸

No path is too far for it to tread; no thought is too subtle for it
　　　to comprehend.

配霑潤於雲雨　象變化乎鬼神

It is a match for clouds and rain in yielding sweet moisture; it is
　　　like spirits and ghosts in bringing about metamorphoses.

被金石而德廣　流管絃而日新

It inscribes bronze and marble, to make virtue known; it breathes
　　　through flutes and strings, and is new always.

APPENDIX I: RHYME SCHEME

Now that I have sunk several craters in the body of the text, I must try
to negotiate peace with the shade of our poet: my plea is that the fissures I
have made in his rhymeprose are strictly metrical and not poetical.

By translating *fu* as "rhymeprose" I have assumed that it is a variety
of prose. Yet I am aware that much controversy has raged over the exact
nature of this genre. For those of the critics who bifurcate all writings into
rhymed and unrhymed classes, *fu* is verse; for those who posit regular rhythmic
patterns as a criterion for verse, *fu* is considered prose. Pending a detailed
study of *fu* rhythms, we may be permitted to take it as prose.

Meanwhile I shall here discuss the rhyme scheme of the *Wên-fu* on the
basis of *Ch'ieh-yün* phonology. It is true that *Ch'ieh-yün* sounds are not
exactly contemporaneous with the date of the *Wên-fu*; but as none of the
rhyme-books of Lu Chi's time is extant, we have to fall back upon the T'ang-
Sung rhyme-patterns.

22

The four preceding translators could profitably have paid a bit more attention to the rhyme schemes of the text. I find all of them lumping together §§ 101-110, which distinctly fall into two heterogeneous parts (101-106 *and* 107-110); one of them even subdivides the whole into two sections, 101-103 *and* 104-110. It is in accordance with rhyme that I have made two sections out of the ten distichs.

I believe I have made profitable use of the *Kuang-yün shêng-hsi* 廣韻聲系, edited by the late Professor SHÊN Chien-shih 沈兼士 of Peking and published in 1945, as well as the detailed study of the rhymes of early Chinese writers by YÜ Hai-yen 于海晏 (漢魏六朝韻譜, in three volumes, Peiping, 1936)—in particular the second volume dealing with the Wei-Chin-Sung-Ch'i writers.

The alphabetization of ancient sounds, worked out by Professor Bernhard KARLGREN, could have been a bit more accurate: e. g., the bilabials b', p, p', m (奉,非,敷,微) might have been distinguished from their labial counterparts (並,幫,滂,明); the same applies to the dental n (泥) versus the palatal n (娘). Finer distinctions will be in demand when alliteration is studied. For our present undertaking, however, they are not of much consequence; LU Chi himself does not seem to be very particular about homophonous rhymes (e. g., see § 2 and § 5).

A		Initials	Dominant Rhyme
§ 1.	墳	b'	
§ 2.	紛	p'	
§ 3.	春 *	ts'	
§ 4.	雲	j	$_{\circ}iu\partial n$
§ 5.	芬	p'	(shang-p'ing 20: 文 $_{\circ}mi u\partial n$)
§ 6.	彬 **	p	
§ 7.	文	m	

* Pronounced $_{\circ}ts'iu\breve{e}n$, this character properly belongs to shang-p'ing 18 諄, $_{\circ}ts'iu\breve{e}n$, a rhyme which does not exist in the *Ch'ieh-yün*, where this "closed" (i. e., rounded, ho-k'o) rhyme is incorporated into the "open" (i. e., not-rounded, k'ai-k'ou) shang-p'ing 17 眞, $_{\circ}tsi\breve{e}n$.

** If it is pronounced $_{\circ}pi\breve{e}n$, this character belongs to shang-p'ing 17. According to *Chi-yün*, however, it can also belong to the dominant rhyme of this section, 文; for there its variant form 份, $_{\circ}p'iu\partial n$ (gloss: 文質備也) is listed under the latter rhyme.

(Cf. Yü 2.34b; also 1.1a, "General Survey," for the indiscriminate rhyming of 眞,諄,文, etc.)

23

B		Initials	Dominant Rhymes
§ 8.	訊 **	*s*	
§ 9.	刵 *	*ᶜńż*	
§ 10.	進 **	*tś*	
§ 11.	潤	*ńż*	*iĕn°*
§ 12.	浸 †	*tsʻi̯əm°*	(ch'ü-shêng 21: 震 *tśi̯ĕn°*)
§ 13.	峻	*s*	
§ 14.	韻 **	*j*	*i̯uĕn°*
§ 15.	振 *	*tś*	(ch'ü-shêng 22: 稕 *tśi̯uĕn°*)
§ 16.	瞬	*ś*	

* 刵振 properly belong to ch'ü-shêng 21 震, *tśi̯ĕn°*.

** 訊進韻, all listed under 稕 in the *Chi-yün*, are to be found under 震,震,問 (*mi̯uən°*, ch'ü-shêng 23) respectively.

As a matter of fact, the 稕 rhyme does not exist in the *Ch'ieh-yün*, where this "closed" rhyme is included in the "open" 震. Strictly speaking, then, the dominant rhyme of this section is *i̯(u)ĕn°*.

† Pronounced *tsi̯əm°*, this character from ch'ü-shêng 52 沁, *tsʻi̯əm°* does not rhyme at all with the other rhyme words of this section. I am almost tempted to emend the text here and alter the character to 汛, *si̯ĕn°* (*Kuang-yün: s. v.* 震), *si̯uĕn°* (*Chi-yün: s.v.* 稕); except that 汛 means (alas) "to sprinkle water on the ground [preparatory to sweeping]." Furthermore, it must not be forgotten that during Han times certain characters ending in -n used to rhyme with 侵, *ᶜtsʻi̯əm* (cf. Yü 1.1b, "General Survey"; for examples see 1.3b-6a, "Table"). This was true only of p'ing-shêng characters, but it is possible that Lu Chi extended this usage and made 浸 rhyme with 震 characters. Incidentally, 浸 can also belong to hsia-p'ing 21, 侵; but it is, of course, in the ch'ü-shêng here.

(Cf. Yü 2.25b.)

C			Rhymes		
§ 17.	班	*ᶜpwan*	shang-p'ing	27	刪
§ 18.	彈	*ᶜdʻân*	" "	25	寒
§ 19.	源	*ᶜngi̯wɒn*	" "	22	元
§ 20.	難	*ᶜnân*	" "	25	
§ 21.	瀾	*ᶜlân*	" "	"	
§ 22.	安	*ᶜân*	" "	"	
§ 23.	言	*ᶜngi̯ɒn*	" "	22	
§ 24.	端	*ᶜtuân*	" "	26	桓
§ 25.	翰	*ᶜɣân*	" "	25	
§ 26.	繁	*ᶜbʻi̯ɒn*	" "	22	
§ 27.	顏	*ᶜngan*	" "	27	
§ 28.	歎	*ᶜtʻân*	" "	25	
§ 29.	然	*ᶜńżi̯än*	hsia-p'ing	2	仙

24

The dominant rhyme here is shang-p'ing 25 $_{\circ}\gamma\hat{a}n$, for it accounts for §§ 18, 20, 21, 22, 25, 28, and § 24 (桓 being nothing but a "closed" variation of 寒). Rhyme 元 is represented by both its classes, "open" (§§ 23, 26) and "closed" (§ 19); the same is true of rhyme 删, "open" (§ 27) and "closed" (§ 17); $_{\circ}\acute{n}\dot{z}\ddot{i}\ddot{a}n$ (§ 29) represents only the "open" class of 仙.

(Cf. Yü 2.25b; for the indiscriminate rhyming of the 元寒桓删山先仙 groups during the Han and San-kuo periods, see 1.1b, "General Survey.")

D		Initials	Rhyme
§ 30.	欽	$k^{\prime}$	
§ 31.	音	$_{\circ}\dot{i}\partial m$	
§ 32.	心	s	$_{\circ}\dot{i}\partial m$
§ 33.	深	$\acute{s}$	(hsia-p'ing 21: 侵 $_{\circ}ts^{\prime}\dot{i}\partial m$)
§ 34.	森	$\underset{.}{s}$	
§ 35.	林	l	

(Cf. Yü 2.41a)

E	(i & ii)	Initials	Dominant Rhyme
§ 36.	量	l	
§ 37.	狀	$d\underset{.}{z}^{\prime}$	
§ 38.	匠	$t\acute{s}$	
§ 39.	讓	$\acute{n}\dot{z}$	
§ 40.	相	s	
§ 41.	當 **	t	
§ 42.	曠 **	$k^{\prime}$	$\dot{i}ang^{\circ}$
§ 43.	亮	l	(ch'ü-shêng 41: 漾 $\dot{i}ang^{\circ}$)
§ 44.	愴	$ts^{\prime}$	
§ 45.	壯	ts	
§ 46.	暢	$\hat{t}^{\prime}$	
§ 47.	誑 *	$k^{\prime}$	
§ 48.	放 *	p	
§ 49.	長	$\hat{d}^{\prime}$	

* Pronounced $k^{\prime}\dot{i}wang^{\circ}$ and $p\dot{i}wang^{\circ}$, these two characters belong to the second, "closed," class of 漾.

** $T\hat{a}ng^{\circ}$ and $k^{\prime}wang^{\circ}$ belong to ch'ü-shêng 42 宕, $d^{\prime}\hat{a}ng^{\circ}$, the former "open" and the latter "closed." Rhyme 漾, however, is but a yodicized variety of rhyme 宕.

(Cf. Yü 2.12a, where 當 is marked as belonging to the 漾 rhyme group.)

F		Initials	Dominant Rhyme
§ 50.	遷	ts'	
§ 51.	妍 **	ng	
§ 52.	宣 *	s	
§ 53.	便	b'	$_{o}i\ddot{a}n$
§ 54.	泉 *	dz'	(hsia-p'ing 2: 仙 $_{o}si\ddot{a}n$)
§ 55.	顛 **	t	
§ 56.	鮮	s	

    * $_{o}Si\underset{.}{w}\ddot{a}n$ and $_{o}dz'i\underset{.}{w}\ddot{a}n$ belong to the second, " closed," class of rhyme 仙.
    ** $_{o}Ngien$ and $_{o}tien$ belong to the first, " open," class of hsia-p'ing 1 先, $_{o}sien$.
    (Cf. Yü 2.28b.)

G		Initials	Dominant Rhyme
§ 57.	章	$t\acute{s}$	
§ 58.	妨 *	p'	
§ 59.	傷	$\acute{s}$	$_{o}iang$
§ 60.	芒 *	m	(hsia-p'ing 10: 陽 $_{o}iang$)
§ 61.	當 **	t	

    * According to KARLGREN's *Analytic Dictionary* (nos. 25 $_{(}p'i\underset{.}{w}ang$ and 1299 $_{(}mi\underset{.}{w}ang)$ these two characters should belong to the second, " closed," class of the 陽 rhyme; SHĒN Chien-shih's edition of *Kuang-yün*, on the other hand, takes them to be of the first, " open," class (p. 529, $_{o}mi\underset{.}{a}ng$; p. 459, $_{o}p'i\underset{.}{a}ng$).
    ** $T\hat{a}ng$ belongs to the first, " open," class of hsia-p'ing 11 唐, $_{o}d'\hat{a}ng$.
    (Cf. Yü 2.8b.)

H			Rhymes		
§ 62.	適	$\acute{z}i\ddot{a}k_{o}$	ju-shêng	22	昔
§ 63.	益	$i\ddot{a}k_{o}$	" "	"	
§ 64.	策	$t\underset{.}{s}'\varepsilon k_{o}$	" "	21	麥
§ 65.	績	$tsiek_{o}$	" "	23	錫
§ 66.	易	$i\ddot{a}k_{o}$	" "	22	

    Each of these three rhymes consists of two classes: " open " and " closed." Here the first class only is employed.
    (Cf. Yü 2.86a.)

I		Initials	Dominant Rhyme
§ 67.	眠	m	
§ 68.	絃	γ	
§ 69.	篇 *	$_{\circ}p'i\ddot{a}n$	$_{\circ}ien$
§ 70.	先	s	(hsia-p'ing 1: 先 $_{\circ}sien$)
§ 71.	捐 *	$_{\circ}iw\ddot{a}n$	

* These two belong to hsia-p'ing 2 仙, $_{\circ}si\ddot{a}n$, one "open" and the other "closed." The three characters of the 先 rhyme are all "open" and not "closed." (Cf. Yü 2.28b.)

J			Rhymes		
§ 72.	致	$\hat{t}i^{\circ}$	ch'ü-shêng	6	至
§ 73.	係	$kiei^{\circ}$	" "	12	霽
§ 74.	緯	$jwei^{\circ}$	" "	8	未
§ 75.	揥 *	$ti\hat{ei}^{\circ}$	" "	12	
§ 76.	媚 **	mji° $(mjwi^{\circ})$	" "	6	
§ 77.	翠	$ts'wi^{\circ}$	" "	6	
§ 78.	偉 †	$jw\hat{ei}^{\circ}$	" "	8	

* In accordance with *Chi-yün*. Both this character and 係 are "open." For a detailed discussion see Appendix IV, Textual Notes.

** According to *Kuang-yün shêng-hsi* (p. 918), this character is "open" (as in 致). KARLGREN (*Analytic Dictionary* no. 608) makes it "closed" (as in 翠).

† According to *Chi-yün*, 偉 is a homophone of 緯, both "closed"; in *Kuang-yün* it is pronounced $^{\circ}jwei$ (shang-shêng 7 尾, "closed" class). Since the word is verbalized, the *Chi-yün* entry may not be incorrect.

K			
(i)		Initials	Rhyme
§ 79.	興	χ	$_{\circ}i\mathschwa ng$
§ 80.	承	$\acute{z}$	(hsia-p'ing 16: 蒸 $_{\circ}t\acute{s}i\mathschwa ng$)
§ 81.	應	$(_{\circ}\cdot i\mathschwa ng)$	

(Cf. Yü 2.21b.)

(ii)			Rhymes		
§ 82.	華	$_{\circ}\chi wa$	hsia-p'ing	9	麻
§ 83.	瑕	$_{\circ}\gamma a$	" "	"	
§ 84.	和	$_{\circ}\gamma u\hat{a}$	" "	8	戈

Of the two characters of the 麻 rhyme one is "open" and the other "closed"; the third, yodicized, is not represented here. 和 belongs to the

27

" closed " variety of 戈, the other two being unrepresented here. Incidentally, 戈 is a rhyme not found in *Ch'ieh-yün*, where it is incorporated into the preceding 歌 rhyme.

(Cf. Yü 2.79a.)

(iii)			Rhymes		
§ 85.	微	ₒ*mjwẹi*	shang-p'ing	8	微
§ 86.	歸	ₒ*kjwẹi*	" "		"
§ 87.	悲	ₒ*pjwi*	" "	6	脂

The 微 characters are here both " closed "; so is the 脂 character.
(Cf. Yü 2.74a.)

(iv)			Rhyme
§ 88.	冶	°*ịa*	°*a*
§ 89.	下	°*γa*	(shang-shêng 35: 馬 °*ma*)
§ 90.	雅	°*nga*	

Of the three classes of the 馬 rhyme, the " open " (§§ 89, 90) and the yodicized (§ 88) are represented here. (The third is " closed.")
(Cf. Yü 2.80b.)

(v)					
§ 91.	濫	*lâm*°	ch'ü-sheng	54	闞
§ 92.	汜	*p'ịwɒm*°	" "	60	梵
§ 93.	豔	*ịäm*°	" "	55	豔

(Cf. Yü 2.43b)

L		Initials	Rhymes	
§ 94.	形	ₒ*γieng*	hsia-p'ing	15 青
§ 95.	情	*dz'*		
§ 96.	輕	*k'*	ₒ*ịäng*	
§ 97.	清	*ts'*	(hsia-p'ing 14: 清 ₒ*ts'ịäng*)	
§ 98.	精	*ts*		
§ 99.	聲	*ś*		
§ 100.	明	ₒ*mịɒng* (ₒ*mịwɒng*)	hsia-p'ing 12 庚	

The character from rhyme 12 represents one of the four subdivisions of that rhyme; " open " according to *Kuang-yün shêng-hsi* (p. 536), " closed " in KARLGREN's *Analytic Dictionary* (no. 634). Rhymes 14 and 15 are each composed of two subdivisions—" open " and " closed "; here we have only the " open " class.

(Cf. Yü 2.16a.)

28

M   Rhyme

§ 101. 服 $b'\underset{\sim}{i}uk_\circ$
§ 102. 淑 $\acute{z}\underset{\sim}{i}uk_\circ$
§ 103. 目 $m\underset{\sim}{i}uk_\circ$ $\underset{\sim}{i}uk_\circ$ (second class—yodicized)
§ 104. 菽 $\acute{s}\underset{\sim}{i}uk_\circ$ (ju-shêng 1: 屋 $\cdot uk_\circ$)
§ 105. 育 $\underset{\sim}{i}uk_\circ$
§ 106. 掬 $k\underset{\sim}{i}uk_\circ$

  (Cf. Yü 2.82a.)

N   Initials   Rhyme

§ 107. 屬 $\acute{z}$
§ 108. 曲 k' $\underset{\sim}{i}wok$
§ 109. 足 ts (ju-shêng 3: 燭 $t\acute{s}\underset{\sim}{i}wok_\circ$)
§ 110. 玉 ng

  (Cf. Yü 2.83a.)

O (i)   Rhyme

§ 111. 紀 $^\circ kji$
§ 112. 止 $^\circ t\acute{s}i$
§ 113. 起 $^\circ k'ji$
§ 114. 理 $^\circ lji$ $^\circ (j)\,i$
§ 115. 齒 $^\circ t\acute{s}'i$ (shang-shêng 6: 止 $^\circ t\acute{s}i$)
§ 116. 擬 $^\circ ngji$
§ 117. 耳 $^\circ \acute{n}zi$

  (Cf. Yü 2.68a.)

(ii)   Initials   Rhyme

§ 118. 留 l
§ 119. 流 l
§ 120. 求 g'
§ 121. 抽 $\hat{t}'$ $_\circ\underset{\sim}{i\partial}u)$
§ 122. 尤 j (hsia-p'ing 18: 尤 $_\circ j\underset{\sim}{i\partial}u$)
§ 123. 勠 l
§ 124. 由 $_\circ\underset{\sim}{i\partial}u$

  (Cf. Yü 2.48a.)

29

P		Initials	Dominant Rhyme
§ 125.	因	$_\circ i\breve{e}n$	
§ 126.	津	ts	
§ 127.	人	$ń\acute{z}$	
§ 128.	泯	m	$_\circ i\breve{e}n$
§ 129.	綸 *	l	(shang-p'ing 17: 眞 $_\circ t\acute{s}i\breve{e}n$)
§ 130.	神	$d\acute{z}'$	
§ 131.	新	s	

* *Liuĕn* belongs to shang-p'ing 18 諄 $_\circ t\acute{s}iuĕn$, which can be considered as the "closed" variety of the 眞 rhyme, a variety of which four species ($_\circ jiuən$, $_\circ kiuĕn$, $_\circ k'iuĕn$, $_\circ \cdot iuĕn$) are still found under that rhyme in *Kuang-yün* and *Chi-yün*. *Ch'ieh-yün* does not make a separate rhyme of 諄.

APPENDIX II: EXPLICATORY NOTES

(Believing that the text is on the whole self-explanatory, I have tried to make as few notes as possible.)

PREFACE

§ 5. Essentially a restatement of the Confucian saying in the *Book of Changes* (繫辭,下): 書不盡言言不盡意. (LEGGE, *The Sacred Books of the East* 16.376-7: "The written characters are not the full exponent of speech, and speech is not the full expression of ideas.")

§ 6. The incommensurability supposed to exist between knowledge and action had already found expression in 非知之實難,將在行之 (*Tso-chuan*, Chao 10—LEGGE, *The Chinese Classics* 5.628; "It is not the knowing a thing that is difficult, but it is the acting accordingly," 630a) and in 非知之難,行之難 (*Ssŭ-ma fa* 司馬法, *Ssŭ-pu ts'ung-k'an* ed. 3.2a). The statement 非知之艱,行之惟艱 occurring in a forged chapter of the *Shu* (LEGGE, *Ch. Cl.* 3.258) must have been inspired by one of these two passages. (In *Chou-li chi-shu* 周禮注疏 22.3b, the subcommentary quotes a passage which is identical with the statement supposedly made by FU Yüeh 傅說 尙書傳詩[var. 說]云非知之艱行之惟艱. This line from a putatively lost ode [逸詩] is quoted by CHU Chün-shêng 朱駿聲[尙書古注便讀, Ch'êng-tu, 1935, 3.17b] and WANG K'ai-yün 王闓運[尙書太傳補注 *Ling-ch'ien-ko ts'ung-shu* ed. 6.8a]. But as the "Collation Note" has it, 尙書傳詩云 is an error for ‖ 傅 ‖; in fact, the subcommentary is quoting from the forged Yüeh-ming 說說命 chapters of the *Shang-shu*.)

The Socratic identification of knowledge with action, which became the keynote of post-Renaissance writers and has now become an item in the credo of many Marxists, was seldom affected by Chinese thinkers until the time of WANG Shou-jên 王守仁, commonly known as WANG Yang-ming 陽明, 1472-1528, nor does it seem to have left any lasting impression on the Chinese intellectual world. At any rate, when SUN Yat-sen reversed the ancient tag and propounded his thesis of 知難行易, he was leaving the identity thesis severely alone.

§ 8. This outlandish line jars me. But I am unable to see how else it can be rendered.

§ 9. Allusion to the line 伐柯伐柯,其則不遠 of Ode No. 158 (LEGGE, *Ch. Cl.* 4.240), which is also quoted in the *Doctrine of the Mean* (LEGGE, *Ch. Cl.* 1.393).

TEXT

§ 1. 玄覽 seems to allude to the line 滌除 | |, 能無疵 in *Tao-tê ching*, chapter 10. ("When he has cleansed away the mysterious sights [of his imagination], he can become without a flaw."—LEGGE, *SBE* 39.54.)

The gloss of "Ho-shang kung" 河上公, however, has 心居玄冥之處,覽知萬事,故云 | |, which makes 玄覽 not a thing to be cleansed away but a result or objective of the process of cleansing. (Arthur WALEY seems to follow this interpretation: "Can you wipe and cleanse your vision of the Mystery till all is without blur?"—*The Way and its Power*, 153.)

§ 4. This more or less baffling couplet means, according to LI Shan, the sublime and the pure (高潔).

§ 6. 彬彬 being derived from the Confucian saying 文質 | |, 然後君子 (*Analects* 6.16; LEGGE, *Ch. Cl.* 1.190), "symmetry" is to be understood in the sense of a correct balance between form and content. On the other hand, the adjunct seems to pull us back from such pedantically Confucian interpretation; "symmetry" may, then, be understood in its usual sense of "due proportion." Then, again, if the expression 林府 is to refer to the term, "symmetry" is not the best translation; instead, we might take the term as tantamount in meaning to 彬蔚, "rich and lush," occurring in § 46.

§ 7. 斯文 may faintly hint at the Confucian connotation of the term (*Analects* 9.5; LEGGE, *Ch. Cl.* 1.217).

§ 9. 八極 ＝ 八方之極, extremities of the eight directions (N, S, E, W, and NE, NW, SE, SW).

§ 11. LI Shan explains *liu-i* as the six arts of the *Chou-li* (ceremonies, music, archery, horsemanship, calligraphy, and mathematics); Ho Ch'o, *op. cit.*, takes them to mean the six Confucian arts (the Books of *Odes*, *History*, and *Changes*, *Ceremonies*, *Music*, and the *Spring and Autumn*).

§ 14. With reference to this passage KU Yen-wu 顧炎武(音論, 1.1b, 音韻學叢書 ed.) observes that LU Chi was the first man of letters to speak of 韻 "rhyme" (which I have translated, rather subversively, as "rhythm"). On the other hand, YEN Jo-ch'ü 閻若璩(尚書古文疏證,眷西堂 ed. 5.15a) writes that the first use of this character was a bit earlier than that: Ts'AO Ts'ao (A. D. 155-220) as mentioned in LIU Hsieh's *Wên-hsin tiao-lung*, chapter 34 章句(昔魏武論賦, 嫌於積韻而善於資代) and in *Chin-shu*, chapter 16 (魏武時河南杜夔精識音韻).

§ 20. Like Saul, who sought his father's asses and found a kingdom.

§ 21. The most sensible explanation of this couplet is that given by Ho Ch'o (see

31

note to § 11): "The two lines probably mean that when a main item is obtained, all subsidiary ones come by themselves 二句疑大者得而小者畢至之意).

§ 32. 素 "silk" is here translated as "paper"; so also in § 116.

§ 38. 司契 probably refers to 有德 ‖ in *Tao-tê ching*, chapter 79. ("[So], he who has the attributes [of the Tâo] regards [only] the conditions of the engagement."—Legge, *SBE* 39.121: "For he who has the 'power' of Tao is the Grand Almoner."—Waley, *The Way and Its Power*, 239.)

§ 39. Patterned after Confucius' 當仁不讓於師. (*Analects* 15.35: "When it comes to acting humanely, you need not be so modest about it as to let your teacher take precedence." Cf. Legge, *Ch. Cl.* 1.304; Waley, *The Analects of Confucius*, 200.)

§ 42. 言窮者 would be "a writer who would explore a subject thoroughly" in order to win over the antagonist. But I confess I am quite baffled by this couplet; for it still does not make much sense at this juncture. The commentators all fail us.

§ 43. The ten literary genres discussed in this and four following couplets do not, of course, exhaust the literature of Lu Chi's days, and yet they seem to be most important ones. It is, furthermore, possible to dispute Lu Chi's description of each of these genres: e. g., a P'an Ta-tao 潘大道 has proposed to emend the first line to ‖‖‖ 深婉 ("profoundly meaningful"? "profound and meaningful"?) on the ground that 綺靡 was the prevalent evil of the effeminate age in which Lu Chi lived (see his *Lun-shih* 論詩, Shanghai, 1927, 21).

§ 48. The second line alludes to the Confucian dictum on the design of the three hundred Odes: 思無邪, "Having no depraved thoughts" (Legge, *Ch. Cl.* 1.146).

§ 49. Another Confucian dictum: 辭達而已矣 (Legge, *Ch. Cl.* 1.305), which can be interpreted in a dozen different ways.

At any rate, as Wallace Stevens writes ("Chocorua to Its Neighbor"):

> To say more than human things with human voice,
> That cannot be; to say human things with more
> Than human voice, that, also, cannot be;
> To speak humanly from the height or from the depth
> Of human things, that is acutest speech.

Which is as good an interpretation as any.

§ 50. 屢遷 may be taken in the temporal sense, but none of the commentators supports this interpretation. Moreover the phrase occurs in the *Book of Changes* (繫辭,下): 易之爲書也不遠,爲道也 ‖‖ ... 不可爲曲要,唯變所適 ("The Yi is a book which should not be let slip from the mind. Its method [of teaching] is marked by the frequent changing [of its lines] . . . , so that an invariable and compendious rule cannot be derived from them;—it must vary as their changes indicate."—Legge, *SBE* 16.399.)

§ 53. The subject "moods" is interpolated.

§ 61. This is translated in accordance with Li Shan's gloss; it seems that the passage should literally mean something like "Whatever is rejected (裁) by your balance deserves (當) to be rejected, even if (固) the things conform to the carpenter's marking-line." Another *Wên-hsüan* commentator, Li Chou-han, seems to take 當 in quite a different sense; his gloss states that "a literary work will conform to the marking-line and (而) becomes exact (相當) if it is tailored by having words and phrases weighed with a balance." This seems to take 其 as equivalent to 而, and 固 as confirmative and not concessive.

§ 68. With regard to 悽, it may be here remarked that a tragic note seems to have prevailed in Chinese poetics since the last days of the Han dynasty; in fact, it seems to have become a frame of reference with which to judge poetry (see 悲 in § 87). As gaiety was a quality not excluded in Confucian poetics (cf. *Analects* 3.20: 關雎 樂而不淫,哀而不傷), it would be worth investigating how and exactly since what time sadness has become the key tune in Chinese poetry.

Is this tearfulness perhaps merely geographical? The elegies of Ch'u are not joyous jingles; could it be, then, that the South has been responsible for the whining note in Chinese poetry?

§ 70. 杼 , "shuttle," is a nice word, for it chimes in with the weaving imagery of § 67. I have, however, translated it as "arrow," a word not foreign to the Occidental literary tradition.

§ 71. For 傷廉 cf. the *Book of Mencius* 4B 23: 可以取可以無取,取 | |, "When to take and not to take are equally correct, you will be impairing your personal integrity by taking."

§ 76. Cf. *Hsün-tzu, Ssŭ-pu ts'ung-k'an* ed. (1)1.11a: 玉在山而草木潤,淵生珠 而崖不枯 , "If there is jade in the mountain, the trees on it will be flourishing; if there are pearls in the pool the banks will not be parched." (Homer H. DUBS, *Hsüntze's Works*, 36).

§ 78. The commentators agree that 下里 and 白雪 were ancient melodies, the former being a sort of jazz tune and the latter an Orphean melody.

§ 79. I like to take the situation described in this and the subsequent couplets as applying to the *haiku*, ancestor of Imagist poetry. Is it possible (I repeat a hackneyed question) to write a long Imagist poem? Can an Imagist draw his breath deep and long?

§ 90. In spite of much controversy that has been raised around 防露 , nothing tangible has emerged out of the fog. At any rate, it must be something not very unlike 桑間 , which is mentioned in *Li-chi* (LEGGE, *SBE* 28.95; COUVREUR, 2.49) and in *Shih-chi* (CHAVANNES, *Mémoires historiques* 3.241).

§ 92. Sacrificial broth was neither salted nor spiced; "vermilion chords" refers to the zithers played in ancestral temples (see the next note).

§ 93. Allusion to 清廟之瑟,朱弦而疏越,一倡而三歎 in *Li-chi* (COUVREUR ed. 2.51).

Notice the crescendo in the five criteria: 應 (§ 81), 和 (§ 84), 悲 (§ 87), 雅 (§ 90), and 豔 (§ 93).

The last term, here rendered as "beauty," properly means "gaudiness." If LU CHI is pleased to pay the highest tribute to an aesthetic standard frowned upon nowadays, it is a case of *de gustibus*. . . .

§ 95. 因宜適變 seems to echo 唯變所適 ; see note to § 50.

§ 100. For the wheelwright P'ien, see *Chuang-tzŭ*, at the end of the chapter 天道 (LEGGE, *SBE* 39.343).

§ 104. 瓊敷 is to be read |王敷. The latter character is listed in *Chi-yün* 平聲,虞 (Ssŭ-ch'uan ed. 2.7a), where 瓊 | is defined as 美玉 , "pretty jade." As 玉藻

(*Li-chi,* Couvreur ed. 1.677) denotes a specific object (beads of jades hanging down from the royal headgear), *ch'iung-fu* should also denote a concrete thing. Is the expression perhaps an error for one of the numerous 瓊 compounds in the *Book of Odes*, all denoting trinkets of one sort or another?

§ 104. Cf. *Book of Odes* (no. 196): 中原有菽,庶民采之 , "In the midst of the plain there is pulse, / And the common people gather it." (Legge, *Ch. Cl.* 4.334.)

§ 105. The couplet can be understood only with reference to *Tao-tê ching,* chapter 5: 天地之間,其猶橐籥乎 , . . . "May not the space between heaven and earth be compared to a bellows? . . ." (Legge, *SBE* 39.50); "Yet Heaven and Earth and all that lies between / Is like a bellows / In that it is empty, but gives a supply that never fails." (Waley, *The Way and its Power,* 147.)

§ 106. The second line refers to 終朝采綠,不盈一掬 in the *Book of Odes* (No. 226), "All the morning I gather the king-grass, / And do not collect enough to fill my hands." (Legge, *Ch. Cl.* 4.411).

§ 107. For "Elevating Discourse" cf. *Shu:* 汝亦昌言,師汝 ‖ (Legge, *Ch. Cl.* 3.76, 78). Needless to say, Lu Chi is not thinking here of Chung-ch'ang T'ung's 仲長統 discourse (*Hou Han-shu, lieh-chuan* 39).

§ 110. For 叩缶 see Li Ssŭ's letter to the First Emperor (李斯,上書秦始皇 in *Wên-hsüan, Ssŭ-pu ts'ung-k'an* ed. [20]39.3b), where he describes how the Ch'in made merry: "Beating water-jars and drumming earthen jugs (扣缶, var. 叩丨), plucking zithers and slapping their shanks, they sing lugubriously to please their ears—this is genuine Ch'in music."

§ 114. 機 is properly a "trigger."

§ 116. For 素 see note to § 32.

§ 118. According to the *Ch'ang-yen* of Chung-ch'ang T'ung (see note to § 107) quoted in Li Shan's commentary, the six emotions are "like" and "dislike" plus the four emotions mentioned in the *Doctrine of the Mean* ("pleasure, anger, sorrow, joy," Legge, *Ch. Cl.* 1.384): 喜怒哀樂好惡謂之六情 .

§ 120. 營魂 may have something to do with ‖ 魄 in *Tao-tê ching,* chapter 10: "the intelligent and animal souls" (Legge, *SBE* 39.53), "the unquiet physical soul," (Waley, *The Way and Its Power,* 153 [Waley's comment: "There is here an allusion to a technique of sexual hygiene "!]). The *Ch'u-tz'ŭ, Ssŭ-pu ts'ung-k'an* ed. (3) 5.7a, also has: 載營魄而登霞兮, in which 魄 is also read 魂 (commentator's textual note). Lu Chi's ‖ 魂, then, may not be a mistake for ‖ 魄 .

§ 121. 若抽 seems to refer to *Chuang-tzŭ* (chapter 天地): 挈水 ‖‖, "It (= the lever) raises the water as quickly as you could do with your hand" (Legge, *SBE* 39.320). Lu Chi, however, meant, it may be assumed, the opposite of "quickly," which is Legge's interpolation.

§ 122. 率意 can mean either "to cudgel the brain" or "to be offhand"; here the expression must mean the latter, just as in the biography of Juan Chi 阮籍 in *Chin-shu* 49: 時 ‖‖ 獨駕不由徑路 . . . , and the biography of Hsiao Fan 蕭範 in *Nan-shih* 52: ‖‖ 題章. The first meaning is to be read into the term as used in the biography of Wang Shao-tsung 王紹宗 in *Hsin T'ang-shu* 199: 常精心 ‖‖, 虛神靜思以取之 and as used in Mei Yao-ch'ên's 梅堯臣 saying quoted in Ou-yang Hsiu's *Liu-i shih hua* 歐陽修,六一詩話 (*Chin-tai pi-shu* ed. 6b): 詩家雖 ‖‖ 而造語亦難.

§ 124. The second half reads 吾未識夫開塞之所由 in the *Ssŭ-pu-ts'ung-k'an Wên-hsüan*, while the other texts omit 也. Even without the final particle, which disturbs the rhyme, the line has nine syllables as against eight in the first half. Yet it is advisable not to delete the innocuous particle 也 or the superfluous connective 之. There is a time and a place for symmetry; § 124 being properly the climax of the body of the text (what follows is anticlimactic), it is not unjustified to think that Lu Chi here broke symmetry intentionally—in order to make the line drag along and sink in the reader's mind. (The construction 吾 . . . 夫 . . . already occurs in § 78.)

§ 125. I think the interpolated line is called for, otherwise 固 would be dangling in the air.

§ 127. The second line is derived from the *Shu*: 予欲觀古人之象, . . . "I wish to see the emblematic figures of the ancients,—the sun, the moon, the stars, the mountains, the dragon, and the flowery fowl, which are depicted *on the upper garment, . . .*" (LEGGE, *Ch. Cl.* 3.80).

§ 128. The first line alludes to *Analects* 19.22: 文武之道未墜於地, "The doctrines of Wăn and Wu have not yet fallen to the ground." (LEGGE, *Ch. Cl.* 1.346.)

§ 129. The *Book of Changes* has (繫辭,上): 易與天地準,故彌綸天地之道, "The Yî was made on a principle of accordance with heaven and earth, and shows us therefore, without rent or confusion, the course (of things) in heaven and earth" (LEGGE, *SBE* 16.353). 塗 in Lu Chi's text, therefore, is to be identified with 道 in this passage.

§ 130. The couplet refers to the *Book of Changes*: the first half compares style with the omnipotent Ch'ien 乾 principle, by virtue of which "the clouds move and the rain is distributed," 雲行雨施 (LEGGE, *SBE* 16.213); the second half may allude to a Confucian saying 知變化之道者其知神之所爲乎, "He who knows the method of change and transformation may be said to know what is done by that spiritual (power)." (LEGGE, *SBE* 16.366.)

§ 131. Cf. 日新之謂盛德 in the *Book of Changes* (繫辭,上), "The daily renovation which it produces is what is meant by 'the abundance of its virtue.'" (LEGGE, *SBE* 16.356.) Cf. also the inscription on the bathtub of T'ang: 苟日日新,日日新,又日新, "If you can one day renovate yourself, do so from day to day. Yea, let there be daily renovation." (Legge, *Ch. Cl.* 1.361.)

APPENDIX III: TERMINOLOGICAL NOTES

Since my aim here is not so much to elucidate Lu Chi—"which would be a task for another lifetime"—as to explain my translation, I shall spare myself the task of juggling with Multiple Definition, the necessity of which Professor I. A. RICHARDS has convincingly demonstrated in his Mencius book. All I propose to do is to clarify, if possible, the use of the following terms recurring in our text.

1. *Wên* 文 seems to operate on two levels. When it appears on the lower level, I have rendered it as "style": Preface § 5 (antithesis: 意), §§ 26 and 125 (antithesis: 理), § 44 (antithesis: 質), § 101 (synonym: 辭). In all these instances

the ideogram could have been translated "language" or "words." As a matter of fact, it is translated "language" in § 62 (antitheses: 理 and 意) and in § 17 (antithesis: 晉). And in 闕文 (§ 14), an expression coined by Confucius (*Analects* 15.25: "a blank in the text"—LEGGE), *wên* is rendered as "words," because it is contrasted with 韻.

On the higher level, *wên* seems to have acquired the status of *Aufhebung*, in which *wên* (and its variations 言,辭) is fused with its Confucian antithesis *chih* 質 (and its variations 理, etc.). Hence I have rendered *wên* in the title (and in Preface § 7) as "literature."

The term 文章 occurring in § 6, where it is used synonymously with 麗藻, "great art" (to be equated with 盛藻, "consummate artistry," of Preface § 7) may allude to the *Analects*, where the expression is used once by Tzǔ-kung with reference to Confucius (5.12; LEGGE, *Ch. Cl.* 1.177) and once by Confucius in eulogizing the emperor Yao (8.19; LEGGE, 1.214), in both instances the expression denoting not "literature" but "music, ceremonies, etc." (cf. LEGGE, 1.214, note). It is possible that LU Chi's reverence for literature was so excessive that he simply identified it with the much abused term "culture." Hence I have capitalized "Literature" for *wên-chang*; I could have done the same with "literature" for *wên*, which on the higher level is hardly differentiated from *wên-chang*.

Ssŭ-wên 斯文, "letters" (§ 7), was first used by Confucius (*Analects* 9.5; "this cause of truth,"—LEGGE, *Ch. Cl.* 1.217) in a sense not very different from *wên-chang*; although my rendering of the expression is defensible, we may assume that LU Chi was also thinking of the Confucian overtone (as brought out in LEGGE's translation).

It is quite sensible to consider *wên* in *tzǔ-wên* 玆文 in § 125 as functioning on the lower level, for it is there contrasted with *li* 理 ; and yet, the compound term can also be taken as a synonym of *ssŭ-wên* (玆 = 斯), in which the ideogram must be accredited with a double-level value.

Finally, the ideogram occurring in 屬文 and 作文 (Preface §§ 4 and 7) seems to have a dubious status: is it of the lower or the higher level? I have avoided the issue by translating the expressions with "to write." For *wên* in § 128, see the Explicatory Note.

2. The two characters *yen* 言 and *tz'ŭ* 辭 cannot but occur frequently in a treatise on literature. I have tried to be as consistent in my rendering of these synonymous terms as possible, translating *yen* as "language" and *tz'ŭ* as "words"—especially when they are yoked together in one couplet (as in Preface § 2, Text §§ 58, 86, 96). When they occur independently, they are not always rendered consistently: *yen* appears as "words" (§§ 23, 115), "language" (§ 51), "diction" (§ 79), *tz'ŭ* as "words" (§ 13, 17, 49, 65, 82, 101) and as "language" (§ 33).

The two expressions 爲言 (§ 23) and 遣言 (§ 51) could have been translated "to speak," like *yen* in § 28. *Yen* in § 100 appears as "to explain"; the ideogram had to be translated beyond recognition in § 42, otherwise I despaired of obtaining any sense out of the couplet. In § 11, 羣言, contrasted with its synonym 六藝, seems to refer to the entire corpus of literature; 片言 in § 64 must mean a short phrase or a terse sentence; for 辭達 (§ 107) and 昌言 (§ 49) see the Explicatory Notes.

3. *Li* 理 (a term discussed in some detail by Professor RICHARDS in *Mencius on the Mind*, 15-16) is translated as "reason" or "reasoning" (§§ 26, 49, 58, 62, 96, 121)

and as "ideas" (§ 125). Essentially, it more or less resembles *logos* in the sense of Platonic ὀρθὸς λόγος (*Phaedo* 73A).

Perhaps one should keep in mind that *li* originally means the grain of a piece of jade and *wên* the pattern on it. This antithesis is noticeable in §§ 26, 62, and 125. In § 49 *li* is contrasted with *tz'ŭ* 辭; and in § 96 it seems to mean the same thing as 喻, "conceits," for it is there used as a synonym of that character and is contrasted with *tz'ŭ*; in § 58 it may be taken as equivalent to 義, "ideas"; finally, *li* seems to serve as a synonym of *ssŭ* 思, "thought," in § 121.

Again, *li* in § 26 may be considered as a reinforcement of 質, "matter" or "content," while in § 129 it acquires a metaphysical status, with the same significance as 塗 (= 道), "Way." (The neo-Confucianism of the Sung era was called *li-hsüeh* 理學 and *tao-hsüeh* 道學 indiscriminately.)

In § 114, *li* is a verb, "to bring to order."

4. The dualism of form and substance discussed in the preceding items runs through the entire piece. A bird's-eye view may be had from the following table.

言 — 思 (§§ 23, 28, 33)   辭 — 義 (§ 17)   文 — 理 (§§ 26, 62, 125)
意 (§ 51)   理 (§§ 49, 58, 96)   意 (Preface § 5)
義 (§ 58)   意 (§ 38)   質 (§ 44)
喻 (§ 96)   藻 — 思 (§ 67)

5. The following table of psychological terms should discourage all translators obsessed with the principle of consistency:

心 — "mind" (Preface § 1, Text § 9, 23, 41, 75). Cf. I. A. RICHARDS, *op. cit.*, 33.
— "heart" (§ 4, 32)
— untranslated (§ 103)
志 — "mind" (§ 1)
— "spirit" (§ 4)
— "mood" (§ 118)

Cf. RICHARDS, *ibid.*, 33.

思 — "to think," "thought" (§§ 2, 8, 23, 28, 33, 67, 115, 121)
意 — "idea" (Preface § 5, Text §§ 51, 62, 75, 122)
— "mind" (§ 38)
義 — "idea" (§§ 17, 58); I took the liberty of equating it with 意.
— "probity" (§ 71); Cf. RICHARDS, *ibid.*, 69.
情 — "emotion" (§§ 1, 27, 43, 118, 122).
— "mood" (§ 10)
— "sincerity" (§ 86)
But 微情, "overtones" (§ 95),
世 |, "the world" (§ 102),
其 |, "insight" (Preface § 4).

Cf. RICHARDS, *ibid.*, 13

慮 — "wits" (§ 23); seems to be a synonym of 思.
神 — "psyche" (§ 118).
精爽 — "soul" (§ 120).
營魂 — "soul" (§ 120); see the Explicatory Note.

Appendix IV: Textual Notes

The text here adopted is on the whole that of Li Shan, as printed in the Hu K'o-chia 胡克家 edition of *Wên-hsüan*, in *chüan* 17 of which is found the *Wên-fu*. Besides, I have consulted the following eight texts:

(1) *Wên-hsüan* 17 in the *Ssŭ-pu ts'ung-k'an* (the so-called 六臣本).

(2) *I-wên lei-chü* 藝文類聚 56, Ming ed., where the text is defectively quoted (Preface §§ 1-10, Text §§ 6, 12-3, 20-1, 23, 25-9, 36-40, 50—the first half, §§ 53-6, 61-3, 66, 71-5, 78-93, 100-10, 114, 123, 126-9 being completely omitted).

(3) *Ch'u-hsüeh chi* 初學記 21, Ming ed., where the text is incompletely (§§ 50 *et sqq.* being totally omitted) and defectively quoted (Preface § 1, Text §§ 12-3, 25-9, 36-40 being also omitted).

(4) Lu Chi's Collected Works in the *Ssŭ-pu ts'ung-k'an* (陸士衡文集).

(5) *Wên-ching pi-fu* 文鏡秘府, facsimile reprint of 1930, Kyōto; Volume 南, where the entire text is preceded by 或曰, "Someone writes . . . ."

(6) *T'ai-p'ing yü-lan* 太平御覽 586 and 588, *Ssŭ-pu ts'ung-k'an* ed.

(7) Li Shan's text as quoted in the commentary of the *Ssŭ-pu-ts'ung-k'an Wên-hsüan*.

(8) The text of the Five Commentators (五臣本) as quoted in the commentary of the *Ssŭ-pu-ts'ung-k'an Wên-hsüan*.

On the basis of these nine texts I took the liberty of making some emendations in the reading here and there.

1. In § 5 all texts have 先人. But | 民 is preferable; as Sun Chih-tsu, *Wên-hsüan k'ao-i* 孫志祖,文選考異(讀畫齋叢書) suggests, 民 was altered to 人 by the T'ang (the former character was a T'ang taboo). *Wên-ching pi-fu* has 民.

2. 浸 in § 12 is a black sheep, for it does not rhyme, but I retain it as a sort of beauty spot; see Appendix I.

3. The second half of § 75 originally reads 意徘徊而不能揥, which has one syllable too many, for the second half is in the predominant pattern of six characters to the line. I took the liberty of deleting 能, without materially altering the meaning. I take 揥 *tiei°* here in the sense of 捐, "to forsake," as given in *Chi-yün* (ch'ü-shêng 12, 霽). It occurs twice in the *Shih* (Odes 47 and 105) in the sense of "comb-pin" (Legge, *Ch. Cl.* 4.77, 165) and is pronounced *t'iäi°* (*Kuang-yün*, ch'ü-shêng 13 祭) or *t'iei°* (*Chi-yün*, ch'ü-shêng 12 霽), but this will not do for our context.

The text of the Five Commentators is supposed to read 襬 for 揥 (cf. *Ssu-pu-ts'ung-k'an Wên-hsüan*) but that character is always in the p'ing-shêng or shang-shêng; that is, it does not rhyme here.

Li Shan proposes to follow the *Shuo-wên* interpretation of the character, i. e., "to take" (取也). But *Shuo-wên* does not have 揥; it does have 扌商, however, (cf. *Shuo-wên chieh-tzŭ ku-lin* 12A.5430a), which is identical with 摘 *t'iek₀* given in *Kuang-yün* ju-shêng 23 錫. (*Chi-yün* ch'ü-shêng 62 lists 扌商 *ti°* and explains it as

棄也; but this must be an error.) As Lu Chi's rhyme schemes are on the whole quite strict, I do not see why Li Shan's proposal should be accepted.

There is another reading suggested by Li Shan: 裼, which appears as 衤雷 in *Shuo-wên* (cf. *Shuo-wên chieh-tzǔ ku-lin* 8A.3713b). Lu Tê-ming (*Ching-tien shih-wên*) reports that Han's text had 裼 in place of 裼 in the Mao text (Ode No. 190, Legge, 307); the latter character is located in *Chi-yün* (ch'ü-shêng 12 霽), t'iei°, and means "a swaddling cloth." Li Shan insists, however, that 裼 means 去也; but this is a mistake. 裼 (*Shuo-wên chieh-tzǔ ku-lin* 8A.3740a) sieko (*Kuang-yün* and *Chi-yün*, ju-shêng 23 錫), in the sense of "to strip off the clothes" occurs in Ode No. 78 (Legge, 129) but it is not known to be interchangeable with 裼 or 衤雷. In short, Li Shan seems to be a bit confused here.

掃   may stand as it is.

4. The last character in § 100 is printed in all texts as 精, which is identical with the last ideogram of § 98. In spite of Ku Yen-wu's statement (cf. *Jih-chih lu* 21, *s.v.* 古人不忌重韻), I have taken the liberty of altering it to 明, which is the reading given in the *Wên-ching pu-fu* text.

5. In § 103 I adopt the reading 蚑 in place of the usual 嗤 or 蚩; Hu K'o-chia in his textual notes recommends this reading. Whichever character is adopted, the meaning is identical.

6. 拘景者 (§ 18) and 短韻 (§ 108) are superior to | 暑者 and | 垣.

Appendix V: Textual Variants

TEXT

§ 3. 喜/嘉   *PF.*

§ 4. 懍懍/凜凜   *LC, CH, FC.*

§ 5. 駿/俊   *LC, CH, FC, PF.*

先民/|人   in all texts except *PF.* (民 was a T'ang taboo.)

§ 6. 嘉/加   *CH.*

麗藻/藻麗   *FC, PF.*

§ 9. 精/晶   *PF.*

鶩/鷔   *PF.*

§ 10. 也/此   *LC.*

瞳曨/瞳矓   *LC.*

§ 11. 漱/瀨   *PF.*

§ 14. 探/採   *LC.* 采 *PF.*

世/代   *FC* (世 was a T'ang taboo).

§ 16. First 於/之

§ 18. 抱/藏   *LC.*

景/暑   *H, L.*

叩/仰   *CH.*

畢/必   *CH, PF.*

懷/壞   *PF.*

§ 19. 沿/緣   *LC.*

而/以   *CH.*

§ 20. 之/末   *FC, CH, PF.*

§ 22. 岨峿/齟齬   *LC.* /鉏鋙 *PF.*

§ 26. 幹/榦   *FC.*

§ 28. 巳/以   *PF.*

§ 32. 緜/綿   *CH, PF.*

沛/霈   *LC.*

乎/於   *CH.*

§ 33. 按/案   *CH.*

逾/愈   *WH, PF* (逾 *H, L*).

§ 34. 青/清   *LC, CH, PF.*

§ 35. 焱/飂   *LC, CH, CW.*

竪/起   *PF.*

§ 38. 效/効   *CW, PF.*

§ 40. 員/照   *CW.* /圓 *FC.*

§ 41. 尙/上   *CH.*

§ 42. 達/遠   *CH.*

§ 44. 緜/綿   *LC, CH, PF.*

§ 46. 頲/頲則   *TPYL* 588.

蔚/欎   *CH.*

精/晶   *PF.*

40

§ 47. 曄/爆   *WH, CW* (曄 *LC, CH, FC, PF*).
§ 48. 邪/雅   *CH.*
§ 49. 辭/詞   *CH.*
      冗/宂   *FC.*
§ 53. 之/而   *CW.*
      而/之   *FC.*
§ 54. 識/相   *FC.*
§ 55. 恒/常   *CW* (恒 was a Sung taboo).
      顚/巓   *CW, FC.*
§ 56. 序/叙   *CW.*
      秩序/袟叙   *WH,* /袟序 *L,* /秩叙 *PF.*
§ 57. 逼/偪   *LC, WH* (逼 *L*).
§ 63. 致/全   *CW.*
§ 64. 而/以   *LC, FC.*
§ 65. 績/勣   *FC.*
      兹/必   *PF.*
§ 67. 千/千   *LC, WH* (千 *L, PF*).
§ 68. 炳/爛   *LC* (丙 and its compounds were avoided by the
                    T'ang), /昞 *PF.*
§ 72. 茗/苕   *PF.*
§ 75. 不掃/不能掃   in all texts except *FC*, which has 不能襐.
§ 76. 暉/輝   *PF.*
§ 77. 於/而   *LC.*
§ 78. 里/俚   *FC.*
      亦/亦以   *FC, PF.*
§ 82. 言徒靡/徒言靡 *H.*
§ 85. Second 以/而   *WH.*
§ 86. 歸/頤   *FC.*
§ 87. 幺/緩   *PF.*
§ 89. 聲高/高聲   *H, L.*
§ 91. 而/人   *FC.*
§ 96. 喩/諭   *LC.*
      辭/詞   *CW.*
      朴/樸   *FC, PF.*
§ 98. 研/妍   *LC.*
      精/晶   *PF.*
§ 99. 猶/循   *LC.*
      赴/趁   *LC.*
      絃/弦   *LC.*
§ 100. 亦/故亦   *CW, PF.*
       明/精   in all texts; 明 in *PF.*

§ 101. 余/予   *WH, PF* (余 *L*).

§ 103. 蚨/虫蛍   *FC.*/蛍 *PF.*/蚨 in all other texts.

§ 104. 瓊/瑤   *CW.*

§ 106. 予/手   *CW, FC.*

§ 108. 韻/垣   *CW, H.*

§ 109. 而/以   *PF.*

§ 110. 於/乎   *PF.*

§ 111. 應感/感應   *CW.*

§ 113. 景/影   *LC, PF.*

§ 116. 毫/豪   *LC, PF* (but marginally corrected to 毫)

§ 117. 以/而   *FC.*

§ 119. 兀/元   *LC.*

§ 120. 攬/覽   *WH, LC* (攬 *L*).
        營/榮   *LC.*/莌 *PF.*
        精/晶   *PF.*
        於/而   *WH, LC, PF* (於 *L*).

§ 121. 逾/愈   in all texts (逾 *FC, PF*).
        乙乙/軋軋   *WH, LC, PF* (乙乙 *L*).
        其/而   *LC.*

§ 122. 是以/是故   *LC, WH* (是以 *L*).

§ 124. 所由/||也   *WH* (所由 *L, H, PF*).

§ 125. 之爲用/其|| *PF.*

§ 126. First 而/使   *CW, WH, PF.*

§ 127. 乎/於   *FC, PF.*
        於/乎   *CW.*

§ 128. 不/弗   *H, CW.*

§ 131. 絃/弦   *LC.*

ADDITIONS AND CORRECTIONS

Page 536, line 15: *For* rarified *read* rarefied
           line 25: *For* prolixty *read* prolixity
Page 537, line 14: *For* sense *read* scene
Page 538, line 2: *For* Weight *read* Weigh
Page 553, line 11: *For* iwok *read* iwok

42

THE *FU* OF T'AO CH'IEN

by

JAMES R. HIGHTOWER

THE *FU* OF T'AO CH'IEN

James Robert Hightower
Harvard University

T'AO Ch'ien is famous as the greatest lyric poet of China before the T'ang dynasty, and his poems have frequently been translated.[1] With the exception of " The Return " 歸去來辭, his rare compositions in the *fu* form [2] are less well known than his lyrics, and justly so. My purpose in offering new translations of T'AO Ch'ien's three *fu* is to show how in each of them he was writing in a well-established tradition, and to point out the nature of his achievement in " The Return," where, by subverting the tradition to his own ends, he made a conventional form the vehicle for intensely personal expression.

The only one of these *fu* which is dated is " The Return," written when T'AO Ch'ien was thirty [3] and at the full maturity of his poetic power. The other two are almost certainly earlier; at least they are avowedly written as poetic exercises, variations on established themes, and should be approached by way of the conventions they accept and exploit. Of these two *fu*, " Stilling the Passions " 閑情 is more nearly a stereotype, and it deals with a theme which does not elsewhere appear in T'AO's poetry. The " Lament for Gentlemen Born out of Their Time " 感士不遇 is equally conventional, but is a topic which he treated frequently in his lyrics and which was apparently more congenial. I shall take them up in order, reserving for last " The Return."

T'AO's preface [4] to " Stilling the Passions " defines the nature of his poem and names two of his models:

---

[1] To those listed in *HJAS* 16 (1953) .265-6 should be added the recent publication *The Poems of T'ao Ch'ien* translated by Lily Pao-hu CHANG and Marjorie SINCLAIR (University of Hawaii Press, Honolulu, 1953).

[2] Following the lead of HSIAO T'ung, who created a special category, " *Tz'u*," for this and one other quite dissimilar composition in the *Wen hsüan*, most translators have failed to observe that this too is a *fu*.

[3] According to LU Ch'in-li's chronology, which I am following; cf. *HJAS* 16.266, note 3.

[4] *Ching-chieh hsien-sheng chi* 5.4b-5a. This and all subsequent T'AO Ch'ien references are to T'AO Chu's edition as published by the Chiang-su Shu-chü, referred to as *Works*.

First of all CHANG Heng wrote a *fu* " On Stabilizing the Passions," and Ts'AI Yung one " On Quieting the Passions." They avoided inflated language, aiming chiefly at simplicity. Their compositions begin by giving free expression to their fancies but end on a note of quiet, serving admirably to restrain the undisciplined and passionate nature: they truly further the ends of salutary warning. Since their time, writers in every generation have been inspired to elaborate on the theme, and in the leisure of my retirement I have taken up my brush to write in my turn. Granted that my literary skill leaves something to be desired, I have perhaps not been unfaithful to the idea of those original authors.

Fragments of these two *fu* by CHANG Heng and Ts'AI Yung appear in T'ang encyclopedias and commentaries, along with several others attributed to writers who lived before T'AO Ch'ien. Though none is complete and there is no guarantee of the authenticity of any of them, they will serve to document the tradition in which he clearly states he is writing. I am putting them in chronological sequence.

Stabilizing the Passions [5]
CHANG Heng (78-139)

Ah, the chaste beauty of this alluring woman! [6]

---

[5] 張衡, 定情賦, *Ch'üan Hou-Han wen* (*CHHW*) 53.9b (全上古三代秦漢三國六朝文). I refer to other collections in this series under the following abbreviated titles: *Ch'üan Han wen*: *CHW*; *Ch'üan San-kuo wen*: *CSKW*; *Ch'üan Chin wen*: *CCW*; *Ch'üan Sung wen*: *CSW*. For *Wen hsüan* (*SPTK* ed.), I am using *WH*, and *YTHY* for *Yü-t'ai hsing-yung*.

CHANG Heng dealt with erotic themes elsewhere, in a part of his *fu* " Meditation on Mystery " 思玄 (*WH* 15.17a-b) and his " Seven Stimuli " 七發, of which a fragment is quoted in *CHHW* 55.2b-3a. In the latter he was of course following the convention established in MEI Sheng's " Seven Stimuli," where all the pleasures of the flesh are elaborately described to distract the ailing prince.

[6] This is the stock opening line of these *fu*, and for convenience I shall bring them all together here, taking the phraseology of CHANG Heng as the standard. Thus, a bar indicates the same character in the corresponding position of CHANG Heng's line.

夫何妖女之淑麗	CHANG Heng
｜｜神｜｜姣｜	SUNG Yü (*WH* 19.9b)
｜｜姝妖｜媛女	Ts'AI Yung (*CHHW* 69.4b)
｜｜淑｜｜佳｜	JUAN Yü (*CHHW* 93.1a)
｜｜英媛｜麗女	WANG Ts'an (*CHHW* 90.2b)
｜｜媛｜｜殊｜	YING Yang (*CHHW* 42.1b)
｜｜美｜｜嫻妖	Ts'AO Chih (*CSKW* 13.4a)
｜｜瓊逸｜令姿	T'AO Ch'ien (*Works* 5.5a)
惟玄媛｜逸女	YANG Hsiu (*CHHW* 51.1a)
美淑人｜妖｜	CHANG Hua (*CCW* 58.1a)

46

She shines with flowery charms and blooming face.
She is unique among all her contemporaries
She is without a peer among her comrades.

The Complaint:

5   Antares' rays decline,[7] insects sing in the grass,
Deep frost falls, vegetation withers;
Autumn is the season, the time is past:
I am distraught as I think of the lovely one.[8]
I imagine I might be the powder on your face
10 But [9] once soiled by dust its radiance is gone.[10]

The title "Quieting the Passions" 靜情 does not occur among
the surviving fragments of *fu* by Ts'AI Yung, but there are a
couple of passages from one called "Curbing Excess" 檢逸 dealing with the same subject. Presumably this is the one T'AO Ch'ien
was referring to.[11]

---

[7] 大火流: cf. *Shih ching* No. 154: 七月流火 "In the seventh month there is
the declining Fire-star" (B. KARLGREN, *The Book of Odes* 99).

[8] These lines are from *I-wen lei-chü* 藝文類聚 (*IWLC*) ·18.13a. The verses
beginning with the "complaint" are probably out of proper context. The tone is
that of some of the *sao* poems, especially the first of the "Nine Persuasions" 九辯
(*Ch'u tz'u* 8.1b-3b *SPTK* ed.). One of the "Nine Declarations" 九章 is called
思美人 (*CT* 4.28a-31b).

[9] 患: literally "I am grieved that . . . ."

[10] These two lines are twice quoted by LI Shan in his commentary on *WH* 19.16b,
34.29a. It is probable that the conceit was further developed in the lost parts of this
fu; cf. T'AO Ch'ien's version.

[11] *IWLC* 18.13a and *T'ai-p'ing yü-lan* (*TPYL*) 380.6b quote nine lines from a *fu*
by Ts'AI Yung 蔡邕 "On Composing the Original [Nature?]" 協初, which appears
to be related to SUNG Yü's "Goddess" 神女賦:

> When she is nearby
> She resembles the supernatural dragon with shining
>   scales about to fly.
> When she has gone afar
> She is like the Spinning Girl in the Milky Way swept
>   by clouds.
> When she stands
> She is like Green Mountain rising majestically.
> When she moves
> She is like the kingfisher beating his wings (cf.
>   "The Goddess" l. 29);

47

Curbing Excess [12]
Ts'AI Yung (132-192)

Ah, this lovely woman of alluring charms!
Her face is radiant and filled with color.
Within all Heaven's bounds she has no equal,
Throughout a thousand years she is unique.[13]
5 My heart rejoices in her chaste beauty
And I am bound to her in unrequited love.
My feelings are without form and have no master.
My thoughts are undecided and swerve to one side.
By day I give reign to my feelings to display my love.
10 By night I depend on dreams to bring our souls together.[14]
I imagine being the vibrating reed [15] in your mouth,
But the notes are solitary [16] and not worth listening to.[17]

Putting a Stop to Desires [18]
JUAN Yü (?-212)

Ah, the exquisite beauty of this virtuous woman!
Her face glows with radiance.[19]
In a thousand generations she has no peer,
Surpassing ancient and modern, she shines alone.[20]

Among all the flaming beauties she has no master.
Her face is like the bright moon,
Her radiance is like the morning sun;
Her beauty is like the lotus flower,
Her flesh like congealed honey.

[12] *CHHW* 69.4a.
[13] 普天壤其無儷,曠千載而特生: cf. CHANG Heng, lines 3-4: 斷當時而呈美,冠朋匹而無雙.
[14] Quoted in *IWLC* 18.13a-b.
[15] 簧鳴 is an inversion; cf. Lu Yün's 陸雲 "Poem on Behalf of Ku Yen-hsien's Wife" (*WH* 25.5a): 鳴簧發丹脣.
[16] I. e., unaccompanied.
[17] 思在口而爲簧鳴,哀聲獨而不敢聆. These two lines are from *Pei-t'ang shu-ch'ao* (*PTSC*) 110.5b. Cf. CHANG Heng, lines 10-11: 思在面爲鉛華兮,患離塵而無光.
[18] 阮瑀,止欲賦, *CHHW* 93.1a-b.
[19] 顏炮炮以流光: cf. Ts'AI Yung, line 2: 顏煒燁而含榮
[20] 歷千代其無匹,超古今而特章: cf. note 13.

48

5 Just now in the first flush of youth
She is both wise and complaisant.
Endowed with the bright virtue of perfect purity
She protects herself by proper conduct.
She would sacrifice her life to do the right thing
10 And so is prepared to emulate Chen-chiang.[21]
My heart delights in her perfect beauty,
Not for a moment do I ever forget her.
I think of the marriage celebrated in the " T'ao-yao " poem [22]
And wish for the shared garment of the " Wu-i." [23]
15 My feelings are all tangled and will not relax.
My soul soars away nine times in one night.[24]
I leave my room and stand uncertain,
I look at the Heavenly River without a bridge; [25]
I sympathize with the Gourd which lacks a mate,[26]
20 I mourn for the Spinning Girl who toils alone.[27]
Then I return to my pillow to seek sleep
That through a dream our souls may meet.[28]
My soul is muddled, it is hard to find hers,

---

[21] Chen-chiang 貞姜 was the wife of King Chao of Ch'u who refused to leave her room with an emissary sent by the King to warn her of a flood because he had forgotten the proper credentials. She chose to remain where she was and be drowned rather than violate an agreement she had made. The story appears in *Lieh nü chuan* 4.16b. (*SPTK* ed.).

[22] *Shih ching* No. 6/1: 之子于歸,宜其室家 " This young lady goes to her new home. She will order well her chamber and house " (KARLGREN, *op. cit.*, 4).

[23] *Shih ching* No. 133/3: 豈曰無衣、與子同裳 " How can you say you have no clothes? I will share my skirt with you " (KARLGREN, *op. cit.*, 86).

[24] 魂一夕而九翔. This common expression occurs in " The Nine Declarations " (*CT* 4.21b), Ssu-ma Ch'ien's " Letter to Jen An " (*WH* 41.26b), as well as in two other *fu* of the present series (YING Yang, line 40, T'AO Ch'ien, line 35), with only minor variants. The " nine " is of course a " complete " number: " any number of times."

[25] I. e., the Milky Way; a reference to the Spinning Girl legend. Literally it is a ford which is lacking.

[26] The Gourd is a star. Ts'AO Chih in his *fu* " The Spirit of the Lo River " (*WH* 19.18b) also bemoans its solitary state: 歎瓠瓜之無匹; likewise WANG Ts'an in his *fu* " Climbing the Pavilion " 登樓 (*WH* 11.3a-b): 懼‖之徒懸.

[27] See note 25. The Spinning Girl is ubiquitous in Chinese poetry from the *Shih ching* (No. 203/5) on.

[28] 庶通夢而交神: cf. Ts'AI Yung, line 10: 夜託夢以交靈.

My thoughts are tangled and confused.
25 At last the night is past, nor have I seen her;
The rising sun in the east marks the dawn.
I know I will not get her for whom I long
And so I control my feelings in writing this.[29]
I stop and stretch my head to see into the distance
Hoping it may be she, but still it is not.[30]

Stilling Evil Passions [31]
WANG Ts'an (177-217)

Ah, this beautiful woman of blooming loveliness!
Her form is truly lovely and of rare beauty.
Nowhere within the four seas has she an equal,
Surpassing all ages she stands out pre-eminent.[32]
5 She resembles the spring flowers of the *t'ang-ti* tree.[33]
In her young maturity she stays at home.[34]

I regret that the year is drawing to a close,
Grieved to be alone with no one to rely on.
My feelings are conflicting and at cross-purposes,
10 My thoughts are melancholy and most grieved.[35]

---

[29] From *IWLC* 18.14a.

[30] Repeatedly quoted in Lɪ Shan's *WH* commentary: 26.15a, 29.30a, 58.7a. The twenty-eight lines quoted in *IWLC* seem to make a satisfactory poem which contains most of the stock elements of these *fu*, and there is no place in its rhyme-scheme to admit this couplet. Of course twenty-eight lines is very short for a *fu*, and there is room for a long digression (after line 16, where the rhyme changes) of which this couplet could well be a part. I have translated the ambiguous 意謂是而復非 as a reference to the 是耶非耶 of Han Wu-ti; cf. *Han shu* 97A.14a (T'ung-wen ed.).

[31] 王粲,閑邪賦, *CHHW* 90.2b. The title is from the *I ching*.

[32] 橫四海而無仇,超遐世而秀出:cf. notes 13, 20.

[33] 唐棣: variously identified as a kind of plum and as the *amelanchier asiatica*, also a flowering tree. Cf. the stanza quoted in *Analects* 9/30 ‖ 之華,偏其反而, 豈不爾思,室是遠而   "The flowery branch of the wild cherry, / How swiftly it flies back! / It is not that I do not love you; / But your house is far away" (WALEY, *The Analects of Confucius* 145).

[34] I. e., unmarried. The *ju-sheng* rhymes end here, and I suspect there is a hiatus in the text. The unannounced change of subject in the next line is otherwise rather abrupt.

[35] 情紛挐以交橫、意慘悽而增悲: cf. Ts'Aɪ Yung, lines 7-8: 情罔象而無 主,意徙倚而左傾; also JUAN Yü, lines 15, 24: 懷紆結而不暢;思交錯以繽 紛.

How miserable my life is fated to be—
My love frustrated and thwarted.
I cross my empty room and go to my bed,[36]
Intending our souls should meet in a dream.[37]
15 My eyes are wakeful, I cannot sleep,[38]
My heart is miserable and uneasy.[39]

Mountains lie ahead of me and the way is obstructed.[40]

I would like to be the bracelet that binds your arm.[41]

Rectifying the Passions [42]
YING Yang (?-217)

Ah, the unusual beauty of this lovely woman!
Complaisant she is, and wisely understanding.
In response to supernatural harmony her substance was
    formed:
She embodies the lush beauty of the orchid and the purity
    of jasper.
5 Among [beauties of] past time rarely equaled,
In present time none can compare with her.[43]
Like the far-reaching rays emitted by the morning sun [44]
The clear glance pours out from her eyes.
In her blooming beauty she crowns our time
10 And is just as virtuous as that woman of Shen.[45]

---

[36] 衽: lit., "a mat," but used for a bed.
[37] 將取夢以通靈: cf. note 28.
[38] 目炯炯而不寐: cf. "The Distant Wandering" 遠遊 (*CT* 5.1b):
夜耿耿而不寐; also *Shih ching* No. 26/1: 耿耿不寐.
[39] To here quoted in *IWLC* 18.14b.
[40] From LI Shan's com. on *WH* 26.10a.
[41] 願爲環以約腕: from *PTSC* 136.8a, with 居 for 邪 in the title. YEN K'o-chün
is right in calling it a misprint. Cf. note 17.
[42] 應瑒,正情賦, *CHHW* 42.1b-2a.
[43] 方往載其鮮雙、曜來今而無列: cf. notes 13, 20, 32.
[44] 發朝陽之鴻暉: the trope is actually more violent: "She emits the far-
reaching rays of the morning sun." Whether this is the radiance of her beauty or the
effect of her glance (as I have taken it) is not clear.
[45] The "Woman of Shen" 申女 refused to marry the man she was engaged to

In my heart I rejoice in that rare beauty [46]

And long for the joy of being joined with her, but there is
   no way.[47]

Overwhelmed by the fragrant beauty of her modest demeanor

My feelings dance around this woman.

15 My soul flutters and goes on its nightly wandering,

I rejoice that our spirits may be united in a common dream.[48]

In daytime I linger hesitant by the roadside,

At night I toss restless until the dawn.[49]

A cool wind blows from the north across the dark wall [50]

20 A cold breeze crosses the middle court.[51]

I hear the high cry of the wild goose in the clouds,

I view the sparkling rays of the massed constellations.

The light of the Southern Star descends like lightning,[52]

The lonely male bird flies swiftly and alone.

25 I hoped [the bird] might lower its head [53] to send me word,

Alas [the star] speeds past and cannot be overtaken.

It grieves me that the passing bird has no mate,[54]

I am sorry that the flowing light cannot be stayed.

---

because the ritual preparations were imperfect. The *Shih ching* poem (No. 17) is supposed to express her resolution. Cf. *Lieh nü chuan* 4.1b-2a.

[46] 余心嘉夫淑美: cf. Ts'AI Yung, line 5: 余心悅于淑麗.

[47] 願結歡而靡因: i. e., no intermediary. Cf. Ts'AO Chih's "Spirit of the Lo River": 無良媒以接歡 (*WH* 19.17b).

[48] 甘同夢而交神: cf. notes 28, 37.

[49] 宵耿耿而達晨: cf. note 38.

[50] 玄序: I can find no other occurrence of this term.

[51] The text seems to be defective at this point, for 唐 does not rhyme with either the preceeding -n or the following -ei rhymes. *T'ang* may be a misprint, but the term 中｜ occurs in another of YING Yang's *fu*, " The Willow " 楊柳; cf. *CHHW* 42.4a.

[52] When the Southern Star appears, the way to the south is open; cf. *Shih chi* 27.9a (T'ung-wen ed.), Ssu-MA Chen's com. (正義). Perhaps the same idea as in TU Fu's poem (寄高適): " The Southern Star['s rays] fall in the old garden. I know for sure he will meet . . ." (*Works* 19.15b, *SPTK* ed.).

[53] For 首 *IWLC* writes 音, a misprint.

[54] 傷往禽之無隅: for 往 *IWLC* has 住. 隅 is a misprint for 偶; cf. JUAN Yü, line 19: 傷弧瓜之無偶.

Too bad the lucky conjunction [55] is just now past,
30 I regret that my desires are all thwarted.
I pace undecided, lost in thought,
My feelings are pained and distressed.[56]
I return to my lonely room and go to bed without
    undressing [57]
I keep tossing and turning without being able to rest.
35 My soul flies afar, sinking and soaring,
Constantly dwelling on her in whom I rejoice.[58]
I look up at the high building [59] and sigh long
Moved by sad echoes, a lingering moan escapes me.
My breath, floating, leaps up to the cloud-house [60]
My bowels in one evening burn nine times.[61]
I imagine myself to be the bright mirror before her
But once gone. . . .[62]

---

[55] For 伏辰 cf. *Tso chuan* (Hsi 5): 龍尾 | |: "[The star] Wei of the [constellation] Dragon lies hid in the conjunction of the sun and the moon" (LEGGE 146). This is cited as a good omen, and I have paraphrased *fu ch'en* as "lucky conjunction."

[56] Lines 21-32 depend for their effect on an elaborate structure and an involved symbolism. The wild goose is the traditional bearer of a message from an absent friend or lover, and a bird is a symbolical intermediary in the "Li sao." A solitary bird is one without a mate, and so represents the poet frustrated in his effort to marry the beautiful woman. Stars are inaccessible, and so a symbol for the unattainable loved one; they also mark the passage of time. Time appears in two aspects: the poet grows old and there is no end to his sorrow; also, the fleeting opportunity passes irrevocably.

These themes are interwoven. Lines 21-24 involve a sort of chiasmus, bird-star: star-bird. The first couplet implies inaccessibility, the second gives promise of transient opportunity. Lines 25-28 shuffle the symbols into a new sequence, bird-star: bird-star, and assert that the opportunity has passed. Lines 29-30 lament passing time and lost opportunity, while 31-32 deal with the resulting state of mind.

[57] For 假寐 cf. *Shih ching* No. 197/2.

[58] For 所覲 read | 歡 as in CH'EN Lin, line 24.

[59] 崇夏 occurs as the name of a temple, but here it should be the name of a constellation, though I can find no support for that interpretation.

[60] 雲館, parallel to *ch'ung hsia* (see note 59), is analogous to | 屋 in the *Chieh-yü* PAN's *fu* "Lament for Herself" 班婕仔, 自悼 (*CHW* 11.7a).

[61] 腸一夕而九煩: cf. note 24. This much is quoted in *IWLC* 18.14b-15a.

[62] 思在前爲明鏡, 哀旣往于替 . . . . This couplet from *PTSC* 136.2a is lacking one word. Cf. notes 22, 46.

Putting a Stop to Desires [63]
CH'EN LIN (?-217)

Lovely! the surpassing woman who lives to the east of my
    house.[64]
Her beauty outshines the spring flowers,
Her charms surpass those of the woman in the " Shih-jen "
    poem.[65]
In antiquity there were few to equal her
5 Today she is indeed without match.[66]
Truly she is one to benefit a state or bring order to a
    household,
Indeed a proper mate for a prince.[67]

How my feelings do take delight in her!
My desires are overflowing and uncontrolled.
10 At night I am restless and unable to sleep,[68]
By day I push aside my food, forgetting hunger.
I am moved at the " If you love me " of the " Pei-feng " poem,
And admire the going home hand in hand.[69]
May the sun and moon move slowly on their courses [70]

---

[63] 陳琳, 止欲賦, CHHW 92.1a-b.

[64] Both " The Lechery of Master Teng-t'u " (WH 19.12b) and " The Handsome Man " (Ku wen yüan, [SPTK ed.] 3.11b) locate the beautiful seductress in the house next door to the east. Cf. also Mencius 6B/1: " Would you climb over the wall of the house next door to the east and abduct the virgin living there? "

[65] Shih ching No. 57/2:

> Her hands like tender shoots,
> Her skin like congealed lard,
> Her neck like insect larvae,
> Her teeth like melon seeds;
> Cicada head, moth eyebrows.
> Her artful smile is red,
> Her lovely eyes clear and black.

[66] 乃遂古其寡儔, 固當世之無鄰: cf. notes 13, 20, 32, 43.

[67] 實君子之攸嬪: cf. Shih ching No. 189/5: 君子攸寧.

[68] 宵炯炯以不寐: cf. notes 38, 49.

[69] For 此 read 北 with IWLC. The " Pei feng " poem is Shih ching No. 41: 惠而好我, 攜手同歸 " If you are affectionate and love me, I will hold your hand and go home with you " (KARLGREN, op. cit., 27).

[70] 忽日月之徐邁: cf. " Li sao," line 9: 日月忽其不淹 (CT 1.7a). I do not understand the hu.

54

15 So that the leafless poplar may put forth sprouts.[71]

I would like to speak to the swallow,
But the swallow flies away, darting up and down.[72]
The way is distant, the road is blocked,[73]
The River is broad and deep, there is no bridge.[74]
20 I stand on tiptoe, wishing to advance,
But it is not a river than can be crossed on a reed.[75]
I loosen the reins and go back home,
Filled with grief, I go to my couch.
Without undressing I close my eyes,[76] and seem at once
　　to sleep,
25 I dream that I see her in whom I rejoice walking toward me.
My soul floats away to my far-off love,
As though we were united and our spirits mingled.[77]

Stilling Thoughts of Love [78]
Ts'ao Chih (192-232)

Ah, the elegant charms of this beautiful woman!

---

[71] I. e., that the poet may still in his old age get a young bride; cf. *I ching* No. 28 (九三): 枯楊生稊，老夫得其女妻 "The leafless poplar puts forth sprouts; an old man gets his bride."

[72] This couplet, quoted by Lɪ Shan (*WH* 31.12a) is inserted here by Yᴇɴ K'o-chün. It would fit the rhyme-scheme better if it followed line 10 (支 rhymes), but it makes a little better sense in the present context. The idea of a bird intermediary goes back to the "Li sao"; in these poems it is usually the wild goose (cf. Yɪɴɢ Yang, line 25) or the phoenix (T'ᴀᴏ Ch'ien, line 31), and the term 玄鳥 has been identified with all three birds. For the swallow and marriage, cf. *Li chi* (*SPTK* ed.) 5.4b: 是月 也玄鳥至 . . . .

[73] 道攸長而路阻: cf. Wᴀɴɢ Ts'an, line 17: 關山介而阻險; also "The Nine Persuasions " (*CT* 8.11a): 路壅絕而不通.

[74] 河廣瀁而無梁: cf. *ibid.*, (*CT* 8.8a): 關梁閉而不通; also the "Lament for the Untimely Fate " (*CT* 14.2a): 江河廣而無梁.

[75] 非一葦之可航: cf. *Shih ching* No. 61/1: 誰謂河廣，一葦杭之 "Who says the River is broad? A single reed crosses it." This is a frequent source of allusion, e. g., Hsɪ K'ang's "Verses for his Elder Brother," No. 9, *Hsi Chung-san chi* 1.2b (*SPTK* ed.).

[76] I have translated 假瞑 as analogous to the *chia mei* of note 57.

[77] 魂翩翩以遙懷，若交好而通靈魂 : cf. Yɪɴɢ Yang, line 15: 翩翩而夕 遊, and notes 28, 37, 48. Except for lines 15-16 (note 72) this passage is from *IWLC* 18.13b.

55

Her rosy face shines with limpid light.
Surpassing and unique, she is without a peer,[79]
So outstanding, in truth, none can equal her.
5 By nature perspicacious and intelligent [80]
In conduct gracious [81] and charming.

I hide where the high peak obscures the sun,
I stand beside the pure current of the limpid stream.
The autumn wind rises in the woods,[82]
10 Lost birds cry as they seek their mates.
Melancholy and laden with grief, my sorrow is the greater,
How can I go on like this? [83]

This by no means exhausts the list of *fu* written before T'AO Ch'ien's time on the subject of stilling the passions, but no more than the titles and a line or two survive of FU Hsüan's "Straightening the Passions," [84] CH'ENG-KUNG Sui's "Assuaging the Passions," [85] or YÜAN Shu's "Rectifying the Passions." [86] P'O Ch'in's "Bringing Sorrow to an End" [87] is not dissimilar, but follows a slightly different pattern in the twenty-six lines which survive, and the same is true for the twenty-three lines of CHANG Hua's "Eternal Love," [88] while Juan Ch'i's "Purifying Thoughts of Love" is an effective parody of the whole idea.[88a]

[78] 曹植,靜情賦, *CSKW* 13.4a.
[79] 卓特出而無匹: cf. notes 13, 20, 32, 43, 66.
[80] 性通暢以聰惠: cf. JUAN Yü, line 6:性聰惠以和良.
[81] 女靡密 is not attested elsewhere. The word 女靡 is to be equated with 靡 as in the similar expressions 靡曼，| 麗; cf. CHU Ch'i-feng, *Tz'u t'ung* 辭通 (abbreviated *TT*) 1967.
[82] 中林 takes on no especial overtones from its three *Shih ching* occurrences. Perhaps the association with "spring feelings" in HSIEH T'iao's poem (和何議曹郊遊) is relevent (*Ku shih yüan* 12.5a, *SPPY* ed.): 春心澹容與,挾弋步中林 .
[83] From *IWLC* 18.15a-b.
[84] 傅玄 (217-278), 矯情 *CCW* 45.4b.
[85] 成公綏 (231-273), 慰情 *CCW* 59.4a.
[86] 袁淑 (408-453), 正情 *CSW* 44.1a-b.
[87] 繁欽 (?-218), 弭愁 *CHHW* 92.8a-b. P'o Ch'in also wrote a *shih* "Settling the Passions" 定情詩 (*Ch'üan San-kuo shih* 3.13a-b).
[88] 張華 (232-300), 永懷 *CCW* 88.1a-b.
[88a] 阮籍,清思, *CSKW* 44.10a-11b. It concludes, "If the myriad phenomena of

So far I have yet to quote an integral specimen of a *fu* on this subject, but already its wholly conventional nature should be apparent. As one reads T'AO Ch'ien's version, the impression of *déjà vue* grows with each couplet. It is perhaps going too far to imagine that every line had its prototype in the original complete texts of those *fu* which time has mercifully destroyed or left in hackneyed fragments, but surely T'AO Ch'ien was not striving for originality in his version.[89]

Stilling the Passions

Ah, the precious rare and lovely form
She stands out unique in all the age.[90]
Though hers is a beauty that would overthrow a city[91]
She intends to be known for her virtue.
5 In purity she rivals her sounding pendant jades[92]
In fragrance she vies with the hidden orchid.[93]
She disowns tender feelings[94] among the vulgar
And carries her principles among the high clouds.
She grieves that the morning sun declines to evening[95]
10 That human life is a continual striving.[96]
All alike die within a hundred years

---

the world do not entangle one's heart, / How is a single female worth being in love with?" 既不以萬物累心兮豈一女子之足思．

[89] *Works* 5.5a-7a. Previous translations of this *fu* are: Anna BERNHARDI, " Grosse Ode zur Beruhigung der Leidenschaften," *MSOS* 15 (1912) .105-9. LIN Yutang. " Ode to Beauty," *A Nun of Taishan and other Translations* (Shanghai, The Commercial Press, 1936) 240-6. Dryden Linsley PHELPS and Mary Katherine WILLMOTT, " Ode to Restrain the Passions by T'AO Ch'ien," *Studia Serica* 7 (1948) .55-62. Lily Pao-hu CHANG and Marjorie SINCLAIR, " Ode to Beauty," *op. cit.*, 113-16.

[90] 獨曠世以秀羣: cf. notes 13, 20, 32, 43, 66, 79.

[91] A reference to LI Yen-nien's song: 一顧傾人城、再顧傾人國 " One glance would overthrow a city, A second glance would overthrow a state " (*Han shu* 97A.13a).

[92] Themselves a symbol of purity; cf. *Li chi* 9.9a.

[93] The 幽蘭 also symbolizes purity; cf. " Li sao," lines 105, 136.

[94] For 柔情 cf. CHANG Hua's *fu* 永懷, line 4: 懷婉娩之 ‖ " In her heart she has the tender feelings of a well brought-up young lady " (*Li chi* 8.26a); cf. also Ts'AO Chih's " Spirit of the Lo River ": ‖ 綽態 (*WH* 19.16b).

[95] Obsession with the passage of time is characteristic of these *fu*; cf. CHANG Heng, lines 6-9; WANG Ts'an, line 7; YING Yang, line 28; CH'EN Lin, line 14; and note 20. It goes back to the " Li sao."

How few our joys, the sorrows how many!
She raises the red curtain [97] and sits straight,
Lightly playing the clear-sounding cither to express her
    feelings.[98]
15 She plays a lovely melody with her slender fingers,
As her white sleeves sweep and sway in time.
A swift glance from her lovely sparkling eyes—
Uncertain whether to speak or smile.[99]

The melody is half played through
20 And the sun is sinking at the western window.
The sad autumn mode [100] echoes through the woods
And white clouds cling to the mountain.
She glances up at Heaven's road,[101]
She looks down and tightens the strings.
25 In spirit and behavior she is charming,
Her attitudes are altogether lovely.[102]

---

[96] 感人生之長勤: cf. "The Distant Wandering," line 10: 哀 ││││ (*CT* 5.2a).

[97] Cf. CHANG Hua's poem 太康六年三月三日後園會 (*Ch'üan Chin shih* 2.4a): 朱幕雲覆 "Red curtains covering like clouds." These are a part of the furnishings of an emperor's boat.

[98] Cf. CHANG Hua's "Love Poem" 情詩 (*Yü-t'ai hsin-yung* 2.10b): 北方有佳人,端坐鼓鳴琴 "In the north there is a beautiful woman, / Who sits straight as she plays her sounding lute." On the proper attitude for playing the lute cf. R. VAN GULIK, "The Lore of the Chinese Lute," *MN* 2 (1939).90, 93. By analogy the cither (*se*) is to be played with the same formality.

[99] 瞬美目以流眄,含言笑而不分. T'AO Chu has 盼 for 眄, presumably because of the *Shih ching* line 美目盼兮 (No. 57/2). However, *liu mien* is a cliché in similar contexts; e.g., CHANG Heng's "Seven Stimuli" (*CHHW* 53.3a): 清眸 ││. For parallels to this couplet cf. "The Lechery of Master Teng-t'u": 含喜微笑,竊視 ││ (*WH* 19.14a); JUAN Chi's "Sorrowful Songs" No. 2: ││ 發媚姿,言笑吐芬芳 (*YTHY* 2.7b). The *pu fen* is not clear. Seitan (*Kokuyaku kambun taisei* 18.264) suggests 忿 "is not angry."

[100] The *shang* 商 mode corresponds to autumn: 孟秋之月...其音 │ (*Li chi* 5.14b). Cf. P'AN Yo's "Lament for a Dead Friend" No. 2: 清商應秋至 (*YTHY* 2.13b): "The clear *shang* mode is in consonance with the autumn season."

[101] 天路 in the many examples cited in *P'ei-wen yün-fu* (*PWYF*) means either "the road to Heaven" or "the Way of Heaven" (天道). In T'AO's poem "The Homing Bird" 歸鳥 (*Works* 1.17a) the term occurs as "a path through the sky" which the bird follows as it navigates through the clouds.

[102] 舉止詳妍: cf. Ts'AO Chih, line 6: 行女靡密而詳妍.

I am moved as she quickens the clear notes' tempo
And wish to speak with her, knee to knee.
I would go in person to exchange vows,
30 But I fear to transgress against the rites.
I would wait for the phoenix to convey my proposal [103]
But I worry that another will anticipate me.[104]
In uncertainty of mind and discomposure [105]
My soul in an instant is nine times transported.[106]

35 I would like to be the collar of your dress [107]
And breathe the lingering fragrance of your flower-
    adorned hair.[108]
But [109] at night you take your silken dress off—
How hateful autumn nights that never end.

I would like to be the girdle of your skirt
40 And bind the modest slender body.
But as weather changes, cool or warm
The old is cast aside, the new put on.

I would like to be the gloss on your hair
As you brush out the dark locks over sloping shoulders.[110]
45 But all too often lovely women wash their hair

---

[103] As in the " Li sao," line 122. Cf. note 47.

[104] 恐他人之我先. Ch'ü Yüan was similarly concerned: 恐高辛之先我 (" Li sao," line 122). The line also appears in Lu Chi's *fu* " On Literature," with 怵 for k'ung.

[105] Cf. WANG Ts'an, line 10.

[106] 魂須臾而九遷: cf. notes 24, 61.

[107] 願在衣而爲領: *lit.*, " I would like to be on your dress, specifically the collar." This same formula is continued in the following stanzas. Cf. notes 17, 41, 62.

[108] 華首 ordinarily means " white hair," but obviously another sense is demanded by the present context. Seitan, *op. cit.*, 267, understands " flower-like face " (華の如き顔); likewise CHU Ch'i-feng (*TT* 2811), but the 餘芳 suggests " hair " as more likely.

[109] 悲 " alas." Here and in the following I have reduced the stock lament to the simple contrast.

[110] �肩. That sloping shoulders were already an attribute of feminine beauty is suggested by the line in Ts'ao Chih's " Spirit of the Lo River ": 肩若削成,腰如約素 " Shoulders as though fashioned by cutting, a waist as though bound by cord " (*WH* 19.16a).

And it is left dry [111] when the water leaves.

I would like to be your penciled eyebrow
To move gracefully with your eyes as you glance around.
But rouge and powder must be fresh applied [112]
50 And it is destroyed as you make up your face anew.[113]

I would like to be the reed that makes your mat [114]
On which you rest your tender body until fall.
But then a robe of fur [115] will take its place:
A year will pass before the mat is used again.[116]

55 I would like to be the silk that makes your slipper
To press your white foot wherever you go.
But there is a time for walking and a time for rest:
The shoes alas are thrown beside your bed.

---

[111] 枯煎 is rather violent for the result of washing, but current shampoo and hair tonic advertisements are quite as extreme in their warnings of what happens to hair washed too frequently without benefit of their panaceas.

[112] 尚鮮 lit., "one esteems freshness [in makeup]." Cf. "A Mistress to Her Lover" 愛姬贈主人 by LIU Hsiao-ch'o 劉孝綽 (cited in *IWLC* 18.8a; cf. *Ch'üan Liang shih* 10.19b): 臥久疑粧脫,鏡中私自看、薄黛銷將盡,疑朱半有殘.
> After lying long abed I suspect my makeup is gone
> And I steal a glance at myself in the mirror;
> My thin-penciled eyebrows have just about disappeared
> And of the rouge only half remains.

[113] 華妝 is makeup for a festive occasion; cf. *Nan Ch'i shu* (T'ung wen ed.) 53.1b: 永明之世,十許年中 ... 都邑之盛,士女富逸,歌聲舞節、袨服 | |,桃花綠水之間,秋月春風之下,蓋以百數 . "During the ten-odd years of the Yung-ming period (483-94), the cities flourished and young men and women were prosperous. Singing and dancing, dressed in their best and with faces carefully made up, hundreds of them disported themselves among flowering peach trees or by the clear streams, under the autumn moon or in the spring breeze."

[114] 願在莞而爲席. Perhaps this is borrowed from CHANG Heng's "Song of Harmony" 同聲歌 (*YTHY* 1.11a): 思爲莞蒻席,在下蔽匡牀 "I imagine myself to be the reed mat / Covering the soft bed beneath you." But here a woman is speaking.

[115] For 文茵 cf. *Shih ching* No. 128/1: | | 暢轂 "Striped floor-mats and protruding wheel-naves" (KARLGREN, *op. cit.* 82). In the present context there can be no question of a floor-mat in a carriage. Cf. the similar use in the CHANG Hua poem quoted in note 97.

[116] 見求: lit., "be sought out."

60

I would like to be your daytime shadow
60 To cleave to your body always, to go east or west.
But tall trees make so much shade
At times, I fear, we could not be together.[117]

I would like to be your nighttime candle
To shine on your jade-like face in your room [118]
65 But with the spreading rays [119] of the rising sun [120]
My light at once goes out, my brilliance eclipsed.

I would like to be the bamboo that makes your fan
To dispense a cooling breeze from your tender hand.
But mornings when the white dew falls
70 I must look at your sleeve [121] from afar.

I would like to be the wood of the *wu-t‘ung* tree
To make the singing lute you hold on your knees
But music, like joy, when most intense turns sad [122]

---

[117] The shadow—lover conceit appears in a poem attributed to Fu Hsüan (217-78) in *YTHY* 9.8a, but to Ch‘e Ts‘ao 車曇攴 (Liang dynasty) in *Yüeh-fu shih-chi* 69.1a, so it is not certain whether the conceit is earlier than T‘ao Ch‘ien. The relevant lines of the poem (車遙遙, a *yüeh-fu* title) are (in the *YTHY* text): 願爲影兮隨君身,君在陰兮影不見,君依光兮妾所願 "I would like to be the shadow that follows your body, / But when you are in the shade your shadow disappears. / That you stay in the light is what I wish."

[118] Besides the occurrence of 兩楹 in ritual contexts (*Li chi* 2.10b, 19.1b) where it means "two pillars"—presumably of the main hall—the expression turns up in one of Ts‘ao Chih's untitled poems 雜詩 (*YTHY* 2.4a): 攬衣出中閨.逍遙步 | | "I take up my clothes and go out of the small gate, / And walk idly between the two pillars." The "two pillars" seem to be outside, but in T‘ao Ch‘ien's *fu* the context calls for an interior scene, specifically a bedroom, though I can find no support for such a metonymy.

[119] For 舒光 cf. "The Goddess" (*WH* 19.9a): 皎若明月 | 其 | "Bright as the full moon spreading its rays."

[120] 扶桑 is metonymy for "rising sun"; the sun rises from the *fu-sang* tree.

[121] Where a fan is kept in hot weather.

[122] 悲樂極以哀來: cf. Fu Hsüan's *yüeh-fu* 明月篇 (*YTHY* 2.9b): 憂喜更相接,樂極還自悲 "Sorrow and joy are close connected, / When joy is most intense one turns sad." The idea is an old one and is quoted as a "saying" in *Shih chi* 126.3a: 酒極則亂,樂極則悲. The pun on music / joy is also well established; cf. the punning definition in *Hsüan-tzu* 14.1a (*SPTK* ed.): 夫樂者樂也 "Music is joy."

And in the end I am pushed aside as you play no more.
75　Put to the test my wishes are all frustrated [123]
And I feel only the desolation of a bitter heart.
Overcome with sadness, and no one to confide in,
I idly walk to the southern wood.[124]
I rest where the dew still hangs on the magnolia [125]
80　And take shelter under the lingering shadows of the
　　　　green pines.
On the chance I should see her as I walk
I am torn in my breast between hope and fear.
To the end all is desolate, no one appears [126]
Left alone with restless thoughts, vainly seeking.
85　Smoothing my light lapel [127] I return to the path
Continually sighing as I watch the setting sun.
With steps uncertain, destination forgotten [128]
Dejected in bearing, face filled with grief.
Leaves leave the branch and flutter down [129]
90　The air is biting as cold comes on.
The sun disappears bearing its rays

---

[123] 考所願而必達: cf. YING Yang, line 30: 哀吾願之多違 ; WANG Ts'an, line 12: 愛兩絕而俱違.

[124] 南林 occurs in Wu-Yüeh ch'un-ch'iu 9.43a (SPTK ed.) as the name of a place where a maiden lives who is an expert with the sword and lance. Usually it is a northern grove 北 | in contexts like the present one; e. g., Ts'AO Chih's yüeh-fu 種葛篇 (YTHY 2.5a): 出門當何顧,徘徊步北林 " Going outside where shall I look? / I walk uncertainly toward the northern grove."

[125] 栖木蘭之遺露: cf. " Li sao," line 33 (CT 1.12b) 朝飲木蘭之墜露 " Mornings I sip the dew hanging from the orchids."

[126] 竟寂寞而無見: cf. " The Distant Wandering " (CT 5.7a): 野寂寞兮無人 " The plain is desolate, no one there."

[127] I am not sure what the significance of this gesture is. 輕裾 usually occurs in contexts where the ch'ing has immediate relevance, as a light garment blown by the breeze (e. g., Ts'AO Chih, 美女篇 YTHY 2.4b). Here it could be intended to suggest poverty—a light garment when the season requires a warm one, but the similar use of the whole expression 斂 | | in a fu " Autumn Sorrow " 秋傷 by CH'U Yüan 褚淵 (Ch'üan Ch'i wen 14.1b) suggests that the ch'ing is simply a part of a cliché, and the gesture itself is equivalent to " with what composure I could summon."

[128] 步徙倚以忘趣: cf. YING Yang, line 31: 步便旋以永思.

[129] 燮燮 occurs in a similar context in one of CHIANG Yen's untitled poems 難體詩 ʽWH 31.22b): | | 涼葉奪. CHU Ch'i-feng (TT 2767) lists a group of similar binoms.

The moon adorns the cloud fringes with light.
With sad cries the solitary bird flies home,[130]
Seeking its mate an animal passes and does not return.

95 I am sorry that the present year is in its decline
I regret that this year draws to a close.[131]
Hoping to follow her in my nighttime dream,[132]
My soul is agitated and finds no rest;[133]
Like a boatman who has lost his oar,

100 Like a cliff-scaler who finds no handhold.

Just now
The winter constellations[134] shine at my window
The north wind blows chill.
I am agitated and unable to sleep,[135]
Obsessed by a host of fancies.

105 I rise and tie my sash to await the morning,
Deep frost glistens on the white steps.[136]
The cock folds his wings and has yet to crow
While from afar floats the shrill sad note of a flute.
At first a harmony of delicate strains,

110 At last it becomes penetrating and sad.[137]

---

[130] 鳥悽聲以孤歸: cf. Ts'ao Chih, line 10 and note 54.

[131] 悼當年之晚暮，恨茲歲之欲殫: cf. Wang Ts'an, line 7: 恨年歲之方暮.

[132] 思宵夢以從之: cf. notes 28, 37, 48, 77.

[133] 神飄颻而不安: cf. Ying Yang, lines 34-5: 固展轉而不安，神耿耿以潛翔; also note 77.

[134] 畢 and 昴 are two fall and winter constellations (Hyades and Pleiades) which rise toward dawn; cf. Ch'en Tzu-ang, "Commemoration of a Banquet at His Excellency Hsieh's Mountain Pavilion" 薛大夫山亭宴序: 東方明而∥升 (Ch'en Po-yü wen chi [SPTK ed.] 7.13b). Cf. also Ssu-ma Hsiang-ju's "Ch'ang-men fu": ∥出於東方 (WH 16.14a).

[135] 怲怲憪不寐: cf. notes 38, 68; also Chuang Chi 莊忌 "Lament for the Untimely Fate" 哀時命 (CT 14.1b): 夜炯炯而不寐. For the variant orthographies of the binom, cf. TT 1555.

[136] Cf. Chang Heng, line 6.

[137] 藏摧. This alliterative binom (anc. *dz'âng dz'uâi*) is used as an onomatopoeia for sad sounds (宛轉何∥∥, of the fulling block in Fei Ch'ang's 費昶 poem in YTHY 6.10a). The same characters in reverse order occur frequently for sad animal cries (e. g., the crane in Chiang Hung's poem in Ch'üan Liang shih 12.11b; the horse in the anonymous "Poem for Chiao Chung-ch'ing's Wife" YTHY 1.19b), and probably form an equivalent term.

I imagine that it is she playing there [138]
Conveying her love by the passing cloud—
The passing cloud departs with never a word,[139]
It is swift in its passing by.[140]
115 Vain it is to grieve myself with longing,
In the end the way is blocked by mountains and crossed
by rivers.[141]
I welcome the fresh wind that blows my ties away
And consign my weakness of will to the receding waves.
I repudiate the meeting in the Man-ts'ao poem [142]
120 And sing the old song of the Shao-nan.[143]
I level all cares and cling to integrity,
Lodge my aspirations [144] at the world's end.[145]

---

[138] The assumption that it is she playing the flute is gratuitous on my part. The line reads 意夫人之在茲.

[139] The conceit originates in the "Ch'ou ssu" 抽思 (CT 4.28b): 願寄言於浮雲 "I wish to send word by the floating cloud." It was used by Hsü Kan in the third of a series of untitled poems 雜詩 (YTHY 1.14a): 浮雲何洋洋,願因通吾辭,飄飄不可寄,徒倚徒相思.

> How vast the floating cloud!
> I would like to use it to convey a message.
> It drifts away before I can send it,
> I vainly think of him in agitation.

[140] 時奄冉而就過: cf. "Li sao," line 9: 日月忽其不淹.

[141] 終阻山而帶河: cf. notes 73, 74.

[142] I. e., 野有蔓草: cf. Shih ching No. 94: "In the open ground there is the creeping grass, the falling dew is plentiful; there is a beautiful person, the clear forehead how beautiful! We met carefree and happy, and so my desire was satisfied." (KARLGREN, op. cit., 61). Cf. CHIANG Yen's fu "Beauty" 麗色 (IWLC 18.16a): 笑月出於陳歌,感蔓草於衛詩.

[143] The "Shao-nan" is the name of a group of poems (Nos. 12-25) in the "Kuo feng" section of the Shih ching. The reference may be to the comments provided by the Preface to the Shih ching on one of these (No. 17): "The manners of a period of decay and disorder were passing away, and the lessons of integrity and sincerity were rising to influence. Oppressive men could not do violence to well-principled women." (LEGGE, The Chinese Classics 4 [Prolegomena] 39).

[144] For 遙情 cf. T'AO Ch'ien's poem "An Outing on the Hsieh Stream" (Works 2.7b): 中觴縱 | |: "The wine half-gone, I give free rein to my aspirations."

[145] 八逷 for the more common | 極 because of the rhyme.

This *fu* of T'ao Ch'ien's is not the last of the series, but there is no point in adding more to the list.[146] Now it is, or should be, a general principle of criticism that an adequate reading of a poem must be based on an understanding of the poet's intent in writing the poem. It has been argued that since the private mental states of the poet are beyond the reach of the critic, all he has to go by is what he finds in the poem he is immediately concerned with, which must be read and judged as something unique. Whatever the theory, good critical practice has never so limited itself. For there are a number of clues to the poet's intent. Sometimes, especially in Chinese poetry, the poet provides a preface to his poem in which he states quite explicitly what he is proposing to do. An intimate knowledge of the poet's life will often suggest attitudes and concerns relevant to understanding a given poem, though such information is usually lacking for Chinese poets. A poet's own statement of his theory of the nature and function of poetry is a valuable guide to his practice. But the most generally available of all these adventitious aids is a knowledge of the poetic tradition in which a poet is writing, and both the genre he is using and the subject of his poem should be viewed in the light of tradition.

I do not propose here to trace the history of the *fu*, a sufficiently complex subject in itself, but shall point out a few features of the form as developed by the Later Han and Six Dynasties periods. Huang-fu Mi (215-282) said,[147] " The *fu* takes its themes from natural objects, whose aspects and properties are elaborated to the point where no one can add anything more." This formula accords well enough with actual practice, and applies both to the descriptive *fu* and, by extension, to the lyric *fu* with which we are presently concerned. Logically such a definition should exclude the possibility of two *fu* on the same subject, for one exhaustive treatment hardly leaves room for a second. However. Ssu-ma Hsiang-ju early established the precedent of taking up a theme already celebrated in a *fu* with the avowed intention of outdoing

---

[146] The most recent seems to be by Hsieh Chi-hsüan 薛季宣 (1134-1173), a *fu* on " Interdicting the Passions "坊情 .

[147] In his introduction to the " Three Capitals " *fu* of Tso Ssu 左思, *WH* 45.40a.

the first effort. With the growth of the popularity of the *fu* this practice was practically the only excuse for writing *fu* at all, as writers became hard put to find new subjects. By early Six Dynasties times not only were the categories exhausted, it was not easy to think of a suitable individual bird, insect, tree, flower, or household utensil that had not been elaborately described in at least one *fu*. Thus as time went on nearly every possible *fu* subject came to be treated in a whole series of *fu*, each member of a series representing a poet's deliberate attempt to incorporate everything his predecessors had written on the subject. This generalization is subject to the usual reservations, but it does apply as a marked tendency that affected the nature of the *fu* form. One result was the production of stereotypes: the development of a subject in any series follows an established sequence, and successive *fu* on that subject differ chiefly in length, the later ones being the longer. In extreme cases even the vocabulary available to the writer of a *fu* on an established theme was to a large degree limited to what his predecessors had used, so that the form is marked by clichés.

At the same time that the *fu* was becoming a stereotyped treatment of a conventional subject, its metrical structure, at one time quite free, was being reduced to a pattern allowing little more variation than the strict *shih* form. From its occasional use as a rhetorical ornament, parallelism became more and more rigid until it was the invariant basis for the construction of each couplet. These various factors combined to make the *fu* little more than an exercise in versification. It was at once a measure of a poet's erudition and an index of his skill if he could write a *fu* to order.[148]

All of these features are abundantly illustrated in the series of *fu* on " Stilling the Passions." Before considering them in detail, the tradition of the subject itself requires a brief treatment. The earliest *fu* containing a catalogue of feminine charms is " The Goddess " (Shen-nü *fu*), of pre-Han date if its attribution to

---

[148] " CHANG Yen, CHANG Shun, and CHU I as youths went to visit CHU Chü, who wished to test them. He said, ' My guests all have to write a *fu* about some object before they can sit down.' Yen wrote on dogs, Shun on mats, I on bows, each writing about something which caught his eye [in the room]. When their *fu* were finished they were seated." *Wen-shih chuan* 文士傳 quoted in *PTSC* 102.3b.

SUNG Yü is accepted.[149] None of the Han Dynasty *fu* on this subject is attested by contemporary mention or quotation, but Ts'AO Chih wrote his "Spirit of the Lo River" (Lo shen *fu*) "inspired by SUNG Yü's description of the goddess for the King of Ch'u," as he said in his preface.[150] Other Han *fu* describe the beauty of some merely human woman and fall into two general types. The one employs a setting where the poet is called upon to disprove a charge of licentiousness; he describes the irresistible temptation to which he was subjected by a lovely and amorous woman whose advances he managed to reject by firmness of will and breath control. Typical examples are "The Lechery of Master Teng-t'u" 登徒子好色賦 and "The Handsome Man" 美人賦, attributed respectively to SUNG Yü and SSU-MA Hsiang-ju.[151] The other type includes the "Stilling the Passions" series, and differs in that the vision of loveliness remains inaccessible. The woman makes no improper advances and so can be praised for her chaste behavior.

The reason for studying these traditions of form and subject is, as I have already suggested, to use them in answering certain fundamental questions about T'AO Ch'ien's poem which must otherwise be obscure. The "Hsien-ch'ing fu" has been variously read as a piece of erotic poetry, as a political allegory, and as a personal love poem. When HSIAO T'ung singled it out among all

---

[149] *WH* 19.8b includes the whole text, but I can find no earlier reference to it than Ts'AO Chih's. In its opening lines the author states that the subject of "The Goddess" is from the "Kao-t'ang" *fu*, which also is attributed to SUNG Yü. In both *fu* SUNG Yü and King Hsiang are characters in the introduction, and it is the character SUNG Yü who is represented as writing the *fu*. This could account for the traditional attribution to the historical person SUNG Yü (of whom exactly nothing is known beyond his supposed association with that king and CH'ü Yüan). I very much doubt that the same man wrote both "The Goddess" and "Kao-t'ang," or that either poem antedates the Han dynasty, but I cannot support my skepticism with facts, and it is convenient to take "The Goddess" as a point of departure.

[150] *WH* 19.15a.

[151] The former is in *WH* 19.12a-14b, the latter in *Ku wen yüan* 3.11a-12b. Neither is mentioned in any text earlier than the sixth century, so far as I know. Very similar to the "Teng-t'u-tzu" is the "Feng fu" 諷, also attributed to Sung Yü in *Ku wen yüan* 2.6a-7b; it is unlikely that both are by the same author. Like those in the *Wen hsüan*, the whole series of *fu* which the *Ku wen yüan* ascribes to Sung Yü (except "The Flute") are about Sung Yü; it is doubtful whether any are by him.

of T'AO Ch'ien's writings as "the one slight flaw on the piece of white jade,"[152] he was presumably indulging in a moralizing judgment.[153] It is likely that his objection was essentially puritanical: a high-minded gentleman like T'AO Ch'ien had no business writing on such a frivolous and questionable subject. This inference is borne out by the fact that HSIAO T'ung excluded from the *Wen hsüan* all the "Palace Style" poems which were being written under the patronage of his brother (HSIAO KANG), though he found room for the occasional pieces, most of them wholly conventional, of his contemporaries. It is not necessary to endorse HSIAO T'ung's critical judgment to agree that he was reading the poem correctly as one of a series of mildly erotic *fu* in which the moralizing twist was not for him a sufficient justification for an unbecoming preoccupation with the more carnal aspects of love.

The allegorical interpretation of the poem is most persuasively stated by LU Ch'in-li.[154] By referring to another tradition, that of the "Li sao," he argues that in both "Stilling the Passions" and the "Spirit of the Lo River" the overt statements of love for a woman really symbolize the love of virtue, and that the poets' melancholy must be understood to be the result, not of frustration in love, but of disappointment of their political ambitions. There is no denying that "Li sao" phrases occur in both *fu*, or that the allegorical tradition is very pervasive in Chinese love poetry generally, but it does not seem to me possible to apply it to the poems in the "Stilling the Passions" series, whatever its validity for the "Spirit of the Lo River," and reading it into the "Goddess" poems involves assuming more than is known about their putative author, SUNG Yü. I quite agree with Mr. LU that the poem should be approached through a study of literary tradition, but it seems to me that he has chosen the wrong tradition.

Finally there is the strictly biographical reading of the poem which insists on taking T'AO Ch'ien's *fu* as the record of a deeply

[152] In his preface to T'AO Ch'ien's *Works* 2b.

[153] His appeal to the authority of YANG Hsiung implies as much, and it was in such terms that SU Shih scolded him (cf. the quotation in T'AO's *Works* 5.7a).

[154] 逯欽立,洛神賦與閑情賦 , *Hsüeh yüan* 學原 2.8 (1948) .87-91.

felt personal experience.[155] Now while there is absolutely no external evidence for attributing to T'ao Ch'ien any such experience, there is at the same time no way of disproving it, and actually the issue is irrelevant to the value of the poem. It is not how deeply the poet feels, but how successfully he persuades his reader to feel. Yet there is a danger, in taking a wholly conventional poem out of its historical context, of accepting a debased currency at its face value. No one expects to find in this sonnet of [156] DRUMMOND's a faithful characterization of the unfortunate Miss Cunningham; he is merely using the established Petrarchian convention of the amatory sonnet: [157]

> The Hyperborean hills, Ceraunus' snow,
> Or Arimaspus (cruel!) first thee bred;
> The Caspian tigers with their milk thee fed,
> And Fauns did human blood on thee bestow;
> Fierce Orithyia's lover in thy bed

---

[155] D. L. PHELPS in *Studia Serica* 7 (1948) .61: "As for this particular ode, one scholastic interpretation has it that the poet's political ambitions met only with frustration, so that finally all he could do was to 'lay his far-reaching feelings to rest in the Eight Horizons'! Thus, the girl his beloved—to these allegorizing scholars— is only the goal of unattainable political ambitions. I do not believe it! The poem is too convincing, too immediate, too direct in sincerity of feeling, for such a dry-as-dust interpretation. I am sure that T'ao Ch'ien was in love, desperately in love, with an irresistible woman! But read the poem for yourself."

[156] Sonnet XXXV, *The Poems of William Drummond of Hawthornden* (*The Muses Library*) 1.69.

[157] The Petrarchizing poets "in sonnet sequence or pastoral eclogue and lyric, told the same tale, set to the same tune. Of the joy of love, the deep contentment of mutual passion, they have little to say . . . , but much of its pains and sorrows—the sorrow of absence, the pain of rejection, the incomparable beauty of the lady and her unwavering cruelty. And they say it in a series of constantly recurring images: of rain and wind, of fire and ice, of storm and warfare; comparisons

> With sun and moon, and earth and sea's rich gems,
> With April's first born flowers and all things rare,
> That heaven's air in this huge rondure hems;

allusions to Venus and Cupid, Cynthia and Apollo, Diana and Actaeon; Alexander weeping that he had no more worlds to conquer, Caesar shedding tears over the head of Pompey; abstractions, such as Love and Fortune, Beauty and Disdain; monsters, like the Phoenix and the Basilisk." (H. J. C. GRIERSON, "John Donne," *The Cambridge History of English Literature* 4.225-6.) This description would need little modification to apply to the *fu* on love themes.

13

> Thee lull'd asleep, where he enrag'd doth blow;
> Thou didst not drink the floods which here do flow,
> But tears, or those by icy Tanais' head.
> Sith thou disdains my love, neglects my grief,
> Laughs at my groans, and still affects my death,
> Of thee, nor heaven, I'll seek no more relief,
> Nor longer entertain this loathsome breath,
> But yield unto my star, that thou mayst prove
> What loss thou hadst in losing such a love.

The sonnet of MARINO (" Te l'Hiperboreo monte, o l'Arimaspe / Produsse, Elpinia, il Caucaso, o 'l Cerauno.") [158] of which DRUMMOND's is not quite a translation, is guarantee of the conventional subject of the poem. In the same way T'AO Ch'ien's *fu* cannot be read in isolation.

The safest point of departure for determining the spirit in which T'AO Ch'ien composed his poem is his preface, where he said in effect that he was writing an exercise on an established theme. It may be worth while to formulate the theme as a preliminary to making a critical estimate of what T'AO Ch'ien did with it. This seems to be the basic structure of the several " Stilling the Passions " *fu*:

There is a woman of great beauty whose equal cannot be found in times past or present. She is good and wise, a model of decorum. I am irresistably attracted to her, but alas! I have no way to approach her. I try to meet her soul in my dreams, but here too I am frustrated. I imagine the bliss of being some inanimate object which she has constantly about her, but realize that all of these are used only to be cast aside. I despair. Finally I resolve to pull myself together, and, by resigning myself to the inevitable, gain some measure of control over my feelings.

T'AO Ch'ien's versification of this formula differs from that of his predecessors—from their surviving fragments, that is—in the considerable elaboration of the metaphysical conceit (lines 35-75), to which approximately one-third of the poem is devoted. The development of each conceit is quite mechanical, and what began

---

[158] Quoted by W. C. WARD on p. 217-8 of *The Poems of William Drummond*.

as a device for relieving monotony becomes monotonous itself. Still this is the section of the poem which most attracts the reader's attention, probably because of the rarity of this trope in Chinese poetry. It would be interesting to know how CHANG Heng, Ts'AI Yung, WANG, Ts'an, and YING Yang used the figure; none of the conceits quoted from their *fu* was borrowed by T'AO Ch'ien, though one at least of his was not original. However, the concept of originality hardly has a place in the critique of pieces like this: success is to be judged according to how well the conventional elements are combined into a harmonious whole. Let us consider T'AO's poem section by section.

The first eighteen lines describe the lovely woman, with the usual emphasis on her moral worth. What at first seems to be an extraneous factor is introduced; she is credited with a mood of melancholy which in the earlier versions was the sole prerogative of the complaining poet. In lines 19-34, the mood is communicated to the poet through the device of the music which she plays, and there is the suggestion that if only a suitable intermediary were available she would welcome his advances (since she too is sad and worried about the passage of time and does not seem to find much consolation in the music). The conceits (" I would like to be the collar of your dress ") in lines 35-74 represent a series of fantasies on how a permanent union might be achieved without the intermediary; in each of them the emphasis is on the irrelevant theme of the impermanence of the imagined propinquity. Instead of saying, " I wish I were . . . but unfortunately that is impossible," he says, " I wish I were . . . but it wouldn't last." This turn appears at first as a welcome deviation from the obvious, but repetition dulls the novelty, and the cumulative effect of this false trail through a third of the poem is to weaken and dissipate its impact. This section does serve to introduce the mood of frustration in lines 75-84, whatever the reason assigned for that mood. The setting of autumn, approaching night, and solitude (lines 85-96), with the usual emphasis on passing time, prepares for the inevitable dream sequence, here condensed into four lines (97-100) and culminating in the effective images of a boat without oars and the cliff-scaler without a handhold. Awake and unable to sleep again, the poet is observing the signs of the night's passing when he hears the sound of a flute (lines 101-110). From the association earlier in the poem of music and his beloved, he is now naturally reminded of her, and again allows himself to imagine that there might be a message from her, brought by a passing cloud rather than the usual wild goose (lines 111-114). Disappointed again and reflecting on the obstacles separating him from the object of his desire, he ends on the note of renunciation and resolution promised in the preface.

In a form which gets its effects by elaboration it is pointless to object to diffuseness, but it is essential that one poem does not develop two unrelated moods. It is in this respect that T'ao Ch'ien's poem is weak. As I have already suggested, the prominence given to the series of conceits is not justified by their contribution to the dominant theme of the poem, that of frustration through inaccessibility.

" Stilling the Passions " is unique among T'ao Ch'ien's works in that none of his usual preoccupations appear in it—added reason for regarding it as an apprentice exercise in versification. The next *fu* which I want to take up is his " Lament for Gentlemen Born Out of their Time." In it are symbols, vocabulary, and above all a theme which he was frequently concerned with. But he is still treating a traditional theme in a traditional manner, as his preface testifies: [159]

TUNG Chung-shu once wrote a *fu* on " Neglected Men of Worth "; SSU-MA Ch'ien likewise wrote one, and as I read them in my leisure time [160] and idle hours [161] I am deeply moved. For to behave that one may be trusted and be concerned to be eligible for Heaven's blessing [162] constitute man's [163] good conduct; to cherish simplicity [164] and maintain equilibrium [165] are the excel-

---

[159] *Works* 5.1a-b.

[160] 三餘之日: cf. *San kuo chih* (*Wei chih*) 13.28b: Asked to explain the meaning of the expression 三餘, [T'UNG] Yü said, " Winter is the idle season 餘 of the year; the night is the idle time of the day; cloudy, rainy days are the idle periods of the weather." Cited by Ho Meng-ch'un 何孟春.

[161] 講習之暇: i.e., " Time when I was not busy carrying on improving conversation with my friends "; cf. *I ching* No. 58 (象): 君子以朋友講習.

[162] 履信思順: cf. *ibid.*, Hsi-tz'u A/11: ｜｜｜乎｜.

[163] 生人 is probably a T'ang emendation to avoid the taboo 民.

[164] 抱朴: cf. *Tao te ching* A.9b: 見素｜｜,少私寡欲.

[165] 守靜: *ibid.*, A.8a: 致虛極,｜｜篤.

lent qualities of the gentleman. Since the time when the true morality departed the world, the great imposture has held sway; [166] in the village they neglect the duty of retiring for high principles, and in the market place they are avid for quick advancement.[167] Worthy men who clung to the right and set their minds on the true way hid their talents [168] in their times; those who kept themselves clean and conducted themselves decently exerted themselves to no purpose to the end of their lives. So Po-i and the [Four] White Heads [169] complained that there was no one to whom they could turn.[170] Ch'ü Yüan [171] gave vent to his cry " It is all over." [172] Alas, we have human form for at the most a hundred years and are gone in the twinkling of an eye; it is hard to establish one's conduct [in so brief a lifetime], but even so the inhabitants of a single city will withhold their unanimous praise.[173] This is why those men of old wet their brushes and repeatedly gave expression to their pent-up feelings without ever resolving them. Now it is only poetry which can give adequate expression to the mind and the feelings. I held a scroll of paper in my hand, uncertain of my powers; finally I was moved to write on this subject.

The rather pedantic tone of the preface suggests that this is a more serious subject. There are only two prototypes, and they are obviously intended to supply background for T'AO Ch'ien's own *fu*. It is hard to tell which was written first, but as it is shorter (the text may of course be incomplete), I shall begin with SSU-MA Ch'ien's:

Lament for Unemployed Gentlemen [174]

Alas for the gentleman born out of his time [175]

---

[166] 大偽: *ibid.*, A.9a. 大道廢有仁義,知慧出則有‖

[167] 易進: cf. *Li chi* 17.7a-b 事君難進而易退則位有序,易進而難退則亂也 " In serving one's prince, when one is reluctant to enter service and quick to retire, then there is order in positions. When one it quick to serve and reluctant to retire, the result is disorder." Cf. also *ibid.*, 9.4b.

[168] 潛玉: For jade as a symbol of a man's worth, cf. *Analects* 9/13.

[169] 夷皓: cf. note 211. The " Four White Heads " were sages who retired from the world under Ch'in Shih-huang-ti and later refused to serve Han Kao-tsu; cf. *Kao shih chuan* B.7a-b (*SPPY* ed.). T'AO Ch'ien elsewhere refers to them in his poetry, e. g., *Works* 2.21a, 3.22a.

[170] Both Po-i and the Four White Heads are credited with similar songs ending with " Whom shall I serve? "

[171] Ch'ü Yüan is addressed by the title 三閭大夫 in " The Fisherman " (*CT* 7.1b).

[172] The expression 已矣 is from the concluding lines of the " Li sao " (*CT* 1.49a).

[173] I am unable to locate the source for this allusion.

[174] 非士不遇: *IWLC* 30.21a-b; *CHW* 26.4b-5a.

[175] For 生之不辰 cf. *Shih ching* No. 257/4: 我生‖ " I was born untimely " (KARLGREN, *op. cit.*, 221).

Ashamed to live alone with only his shadow for companion,[176]
Always concerned to control himself and be courteous [177]
Fearing lest his determination to act go unmarked.[178]
5 In truth his endowment is adequate but the time is out
    of joint;
Endlessly he toils up to the very verge of death.
Though possessed of [pleasing] form, he goes unnoticed,
While capable, he cannot demonstrate his abilities.
How easily people are misled by poverty or success—
10 It is hard for them to distinguish between beauty and
    ugliness.
While time drags on and on
I am hemmed in, never given scope.
He who treats the just justly
    Is my friend:
He who is selfish with the selfish
    Brings grief to himself,[179]
15 Heaven's way is mysterious
    Vast indeed; [180]

---

[176] 顧影 lit., "watching one's shadow," comes to mean "self-absorbed," out of vanity of either worth or beauty.

[177] 克己而復禮: cf. *Analects* 12/1: "To control oneself and be courteous is perfect virtue (*jen*)."

[178] 無聞: cf. *ibid.*, 9/23: "A youth is to be regarded with respect. How do we know his future will not be equal to our present? If he reach the age of forty or fifty, and has not made himself heard of, then indeed he will not be worth being regarded with respect" (LEGGE, *The Chinese Classics* 1.223).

[179] 使公于公者 . . . 私于私者: I do not understand these lines, which perhaps should be referred to *Lieh tzu* (*SPTK* ed.) 1.6b: 公公私私天地之德. A rich man told a poor man he had got his wealth by stealing. The poor man tried it and was arrested. The rich man explained that by stealing he meant stealing from nature, not from men. Master Tung-kuo commented, "Mr. Kuo's stealing was from the common store, and so he escaped punishment. Your stealing was selfish interest, and so you got into trouble. Both those who treat the private as though it were public and those who do not do so are thieves. To regard the public as public and the private as private is the principle of heaven and earth. Knowing the principle of heaven and earth, who will speak of stealing or not stealing?" This passage suggests a possible translation: "He who treats public [property] as public is my friend; he who appropriates what belongs to others brings grief to himself." I am not sure what that would mean in the present context.

[180] 天道微哉、呼嗟闊兮: Li Shan's com. on *WH* 15.26a, 24.13b, 28.17b has ‖ ‖ 悠時,人理足兮. As YEN K'o-chün remarks, this is a contamination from the next line.

The way of the world is obvious:
　　Overthrow and rape.[181]
To love life and hate death
　　Is despised by the able;
To love rank and insult the lowly
　　Is the overthrow of the wise.
Brilliant is my deep insight
　　My understanding capacious.
20 Murky is their unenlightenment
　　Poison brewing within.[182]
This heart of mine—
　　The wise man understands it; [183]
These words of mine—
　　The wise man garners them.
To die nameless
　　Was the ancient's shame; [184]
To hear the truth in the morning and die that night—
　　Who will say the sage was wrong? [185]
25 There is a cycle between bad times and good:
　　[States] fall and rise.
One cannot depend upon constant principles
　　Or rely upon sound knowledge.[186]
Do not act to bring about happiness,

---

[181] For the contrast between 天道 and 人理 cf. *Chuang tzu* (*SPTK* ed.) 4.42b: 何謂道.有天道有人道.無爲而尊者天道也.有爲而累者人道也 . . . 天道之與人道相去遠矣 "What is meant by the Way? There is the Way of Heaven and there is the Way of man. To be esteemed without acting is the Way of Heaven. To become involved through acting is the Way of man. . . . The Way of Heaven and the Way of man are far apart."

[182] 內生毒: there must be an allusion behind this phrase which I have not been able to discover.

[183] 我之心矣,哲巳能忖 : cf. *Shih ching* 198/4: 他人有心,予忖度之 "Other men have their thoughts, but I can understand them " (KARLGREN, *op. cit.*, 148).

[184] This is the theme of the "Letter to Jen An," *WH* 41.10a-27a, esp. 23b: 沒世而文采不表於後世也.

[185] Cf. *Analects* 4/8: "The Master said, 'If a man in the morning hear the right way, he may die in the evening without regret'" (LEGGE, *op. cit.*, 1.168).

[186] These two lines are from Lɪ Shan's com. on *WH* 39.26a. This is a repudiation of the Confucian concepts of 理 and 智, leading to the Taoist conclusion.

Do not interfere to precipitate calamity: [187]
Entrust yourself to the spontaneous
And in the end everything will revert to the One.

In this, Ssu-ma Ch'ien (if the poem is actually his) gives some weight to the old charge of Taoist inclinations, but typically he provides an unimpeachably Confucian setting for his heresy. Tung Chung-shu's treatment of the theme is rather more ambitious:

Neglected Men of Worth [188]

Tung Chung-shu

Oh, alas, how far-off, how distant! [189]
How slowly the chance comes, that so swiftly recedes.[190]
They are no followers of ours [191] who bend their will to
    others' beck; [192]
Upright [193] I have awaited my chance until now I am
    approaching the grave.
5 Time goes on,[194] I cannot expect to be understood,[195]
My heart is depressed,[196] I cannot hope for á position.
Uneasy activity would serve only to add to my disgrace,

---

[187] 無造福先、無觸禍始: cf. *Chuang tzu* 6.3a: 不爲福先、不爲禍始 " He does not take the initiative in producing either happiness or calamity " (Legge, *SBE* 39.265).

[188] 董仲舒,士不遇: *KWY* 3.3a-3b; *IWLC* 30.20b-21a; *CHW* 23.1a-b.

[189] This apostrophe is presumably addressed to Heaven; cf. Ssu-ma Ch'ien, line 15; also the common expressions 天高地遠 and ‖ 星遠. The burden of this plaint is that fate is unknowable.

[190] 時來曷暹,去之速矣: cf. *Fa yen* (*SPTK* ed.) 6.2b: 辰乎辰,曷來之暹 去之速矣 " The good time, the good time, how slowly it comes and how fast it goes."

[191] Cf. *Analects* 11/17. By appropriating to himself Confucius' words the poet is making himself the spokesman for the Confucian tradition.

[192] 屈意從人: cf. Ssu-ma Ch'ien, line 12: 咸 遂屈而不伸.

[193] 正身: Cf. *Mencius* 4A/4: 其身正而天下歸之.

[194] 悠悠時偕: cf. Ssu-ma Ch'ien, line 11: 時悠悠而蕩蕩.

[195] 豈能覺矣: I take the *chüeh* as referring to his potential patron, the ruler who might employ him; cf. the " Biography of Ch'ü Yüan," *Shih chi* 84.3b: 懷王之 終不悟也.

[196] 心之憂歟 (with 矣 for *yü*) occurs frequently in the *Shih ching*, e. g.; No. 26/5.

To butt the fence with all my strength will only break
my horns.[197]
If I do not leave my door I may avoid trouble.[198]

Development:[199]

10 I was born not during the flourishing of the Three Dynasties
But during the time of decadence which followed them.
While through cleverness and deceit one can expect success,
The upright and the uncompromising exercise self-
restraint.
Though I thrice daily reflect on my conduct [200]
15 I am fully aware that to advance or retire is equally
difficult.[201]
Men of that ilk truly there are many [202]
Who point at the white and call it black.[203]
It is pretty eyes which are trusted, but my sight is dim,
Glib tongues are believed, but my speech is faltering.[204]
20 The gods are unable to straighten out the perversity of
human affairs
Nor can sages enlighten the befuddlement of the stupid.
If I leave my door [205] I cannot walk together with them

---

[197] Cf. *I ching* No. 34 (九三): "A ram butts against a fence and entangles his
horns."

[198] *Ibid.*, No. 60 (初九, "He does not leave his door. No blame." This unpaired
line does not end in a rhyme, and either the 過 is a misprint or a line has dropped
out of the text.

[199] 重曰. This term occurs in the "Distant Wandering" (*CT* 5.4a) and in the
Chieh-yü Pan's "Lament for Herself" (*CHW* 11.7a).

[200] Cf. *Analects* 1/4: "I daily examine myself on three points" (LEGGE, *The Chinese
Classics* 1.139).

[201] 進退惟谷: cf. *Shih ching* No. 257/9: "People have a saying, 'To advance or
retire is alike difficult.'"

[202] 實繁之有徒: cf. *Shu ching* 4/2/3: "Contemners of the worthy and parasites
of the powerful,—many such followers he had indeed" (LEGGE, *op. cit.*, 179).

[203] Cf. "The Nine Declarations:" 變白以爲黑 "They transform the white and
make it black" (*CT* 4.25a).

[204] 目信嫮而視眇兮口信辨而言訥. It is not clear just how these attributes
are to be distributed. On the basis of *Analects* 4/24 ("The superior man wishes to be
slow in his speech and earnest in his conduct"), 言訥 should be a positive virtue,
contrasting with "glib-tongued." By analogy the same distinction should hold between
mu hu and *shih miao*.

[205] 出門: i.e., take office.

When I hide my talents [206] they scoff at my intransigence.
I withdraw to cleanse my heart [207] and examine my
    conscience [208]
25   But still they do not understand the course I follow. [209]
When I consider conditions in ancient times,
    Then too men of integrity were isolated and had no
    one to turn to.
Under T'ang of the Yin there were Pien-sui and Wu-kuang [210]
Under Wu of the Chou there were Po-i and Shu-ch'i. [211]
30 Pien-sui and Wu-kuang drowned in the deeps
    Po-i and Shu-ch'i climbed the hill to pick herbs.
If even saints like those were distraught [212]
    What is to be expected when the whole world has gone
    astray?
Men like Wu Yüan [213] and Ch'ü Yüan [214]
35   Were really without anyone they might look to. [215]
Though I am not up to [the conduct of] those men,
    I shall go on a distant voyage, [216] always admiring them.
Alas, men of my sort are far away [217]

---

[206] 藏器: cf. *I ching* (Hsi-tz'u B/4): 君子 ｜ ｜ 於其身,待時而動 "The superior man keeps his instrument concealed on his person, awaiting the proper time to act."

[207] 退洗心: cf. *ibid.*, (Hsi-ts'u A/10): 聖人以此洗心,退藏於密 "The sages with these cleansed their hearts and, retiring, treasured them up in secrecy."

[208] For 內訟 cf. *Analects* 5/27.

[209] 亦不知其所從: This may mean "I do not know to whom I might offer allegience."

[210] T'ang first offered Pien-sui the empire, and Pien-sui drowned himself. He then offered it to Wu-kuang, who likewise drowned himself. Cf. *Chuang tzu* 9.29b-30a.

[211] For Po-i and Shu-ch'i cf. *BD* 1657.

[212] Chu Ch'i-feng (*TT* 878) equates 周逴 with 彷徨 (along with other variant orthographies); the meaning ranges between "idle" and "uncertain."

[213] Better known as Wu Tzu-hsü; cf. *BD* 2358.

[214] *BD* 503.

[215] For protection and employment.

[216] The connotations of 遠遊 are Taoist (*CT* 5) and allegorical of a search for a patron ("Li sao"). As developed in the following lines, it must also be taken literally: "I am going on a voyage of discovery to see whether I can find a sympathetic friend."

[217] In time (as the misunderstood heroes just mentioned) and in space (as the friend he hopes to find.)

I fear the path is overgrown and hard to walk;
40 I dread the warning that the superior man on a journey
　　Will go three days without eating.[218]
Alas, everyone in the world is perverse
　　I regret there is no one to join me in getting back
　　　[to the True Way].
Better turn to the good old cause
45 　And not let oneself be carried along by the times.
Though all profit be gained by violating the true self
　　It is still better with pure heart to cleave to the one Good.
One may act only under pressure of circumstances—
　　It does not follow that he is by nature obstinate.[219]
50 I know well that great achievement [220] comes with
　　　companionship [221]
　　And understand the rewards of the glory of humility.[222]
I conform to the hidden through silent contentment
　　And do not show off my excellence or seek to be prominent.
If one can make common cause with a true friend [223]
55 　Why quibble over the difference in our ages? [224]

Tung Chung-shu has developed the theme by supplying examples, and for consolation looks for a friend in adversity. The
life of retirement which he advocates is well within the Confucian
tradition of staying out of office when the times are bad. T'ao
Ch'ien multiplied the examples and borrowed freely from both his
predecessors:

---

[218] Cf. *I ching* No. 36 (初九): "The superior man on a journey will go three
days without eating."

[219] 紛紜迫而後動兮豈云稟性之惟褊: I am not sure that I understand this
line. I take it to mean that, though unwilling to compromise his ideals by serving
when the times are not right, he might be forced by circumstances (e. g., poverty) to
do so; still his reluctance is not to be taken as a sign of obstinacy, for he would
gladly serve if he could do so on his own terms.

[220] 大有: *I ching* No. 14.

[221] 同人: *ibid.*, No. 13.

[222] 謙 *ibid.*, No. 15 (象): 尊而光.

[223] 肝膽: lit. "liver and gall."

[224] 髮鬢: I have been unable to locate the reference behind this term, and so my
translation is only a guess. It might imply "superficial differences" rather than age.

Lament for Gentlemen Born out of their Time
T'ao Ch'ien

Ah, of all who receive the breath of life from the Creator [225]
It is man alone who is endowed with intelligence.[226]
One, given divine knowledge, hides his light;
Another, possessed of the Three and the Five,[227] leaves
     a name to posterity.
5 Some find their satisfaction in breaking clods,[228]
Others perform some great service to mankind.
Granted that quiescence or activity [229] are allotted by fate,
Whatever the circumstances one should be complacent
     and satisfied.
The world floats along and goes its way,
10 While all things are divided into classes according to form.[230]
When a fine net is cast the fish are frightened,
When a strong snare is laid the birds are alarmed.[231]

---

[225] 大塊: cf. *Chuang tzu* 2.19a: 夫｜｜噫氣,其名爲風 " The breath of the Creator is called the wind."

[226] Cf. *Shu ching* 21/1/3: 惟天地萬物之父母、惟人萬物之靈 " Heaven and Earth are father and mother of all things, and man of all creatures is the one endowed with intelligence."

[227] 三五 is ambiguous. It may refer to the 三皇五帝 (as in PAN KU's " Two Capitals " *fu*, *WH* 1.17b), or two constellations (*Shih ching* No. 21/1), or the significant time intervals of thirty and five hundred years (*Shih chi* 27.37a, 41b). It is the last which yields the best sense here; cf. the first *Shih chi* passage cited: 爲國者必貴三五 " Rulers of a state must respect the three and the five."

[228] 擊壤: " In the time cf the Emperor Yao the world was at peace and the people were at rest. An old laborer, over eighty, was breaking clods in the road. Someone who saw him said, ' Great is the virtuous power of the Emperor!' The worker said, ' I begin work at sunrise and rest at sunset. I dig a well for water and till my field for food. What is the Emperor's virtue to me? ' " (*Kao shih chuan* A.4a). LI Kung-huan's note about a game of darts going under the same name is irrelevant.

[229] 潛躍: i. e., whether one lives in retirement or leads an active life of public service; cf. *I ching* No. 1 (初九).

[230] 物羣分以相形: cf. *ibid.*, (Hsi-tz'u A/1): 物以羣分、吉凶生矣 " Creatures are divided by classes; from this come good and bad fortune." Perhaps there is also a reflection here of the couplet in No. 6 of the poems " On Drinking Wine " (*Works* 3.22) 是非苟相形、雷同共毀譽 " When right and wrong are arbitrarily given form,/They all join together with their blame or praise."

[231] Cf. JUAN Chi's " Sad Songs " No. 76 (59a in HUANG Chieh's ed.): 綸深魚淵潛,矰設鳥高翔 " When the line hangs deep the fish dive into the depths; / When the arrows fly aloft the birds soar high."

In the same way the truly wise are quickly put on their guard
And flee from office to go back to farming.
15 High-soaring mountains hide their shadows,
Broad-flowing rivers conceal their sounds.[232]
They sigh long when they think of the Emperors Huang-ti
and Yao;
Relinquishing glory, they take pleasure in poverty and
low condition.

The water in flowing from the pure spring is forever divided,
20 Through action good and evil take their separate courses.
When we look for the most estimable kind of conduct
It is surely the good in which one can take most pleasure.
We accept our lot from Heaven above
And take as our guide the writings bequeathed by the
Sages.[233]
25 We show ourselves loyal to our prince and filial to our parents
We cultivate trust and duty in our town.
We will gain distinction [if at all] through honesty
Never seeking praise if it involves compromise of principle.[234]

Alas, the sycophants and slanderers—
30 The world abhors anything superior.
The man of vision they call deluded,
The one whose conduct is upright they say is perverse.
He who is absolutely righteous and above suspicion
In the end is put to shame with slanderous charges.
35 You may clasp your jewel and cling to your orchids,[235]

---

[232] 山巖巖而懷影, 川汪汪而藏聲. I take these lines as referring to the
hermit's retreat, for which mountains and rivers are common symbols; cf. *Shih chi*
79.20b: 退而巖居川觀 Cf. also Ts‘ao Chih's "Stilling Thoughts of Love," lines
6-7: 蔭高岑以翳日、臨淥水之清流.

[233] Cf. T‘ao Ch‘ien's poem "To YANG Sung-ling" (*Works* 2.20b): 得知千載事，
正賴古人書 "To know about times a thousand years ago, / We have only to rely
on the writings of the ancients."

[234] Cf. TUNG Chung-shu, lines 16-7.

[235] Symbols of the worth of the upright man. For the jewel cf. note 168; for the
orchid cf. note 93.

In vain your fragrance and purity—who believes in them?

Alas for gentlemen born out of their time!
I can no longer live under Shen-nung or the Emperor K'uei.[236]
In solitude I have devoted myself to self-cultivation— [237]
40 When have I failed thrice daily to examine myself? [238]
I hoped that by improving my virtue I would be ready
    if a chance should come,[239]
The chance came, but I found no favor.
Without a direct word from Master Yüan
Chang Chi would have died in obscurity.[240]
45 I sympathize with Old Man Feng, the Palace Secretary
Who had to depend on Prefect Wei to give his advice.[241]
They made every effort to achieve recognition:
Still they ate their hearts out, year after year.
One may be sure there is no tiger in the market
50 But three reports will lead one astray.[242]
I lament the Tutor Chia's [243] outstanding talents
His far-reaching course checked and confined in bounds.[244]

---

[236] 帝魁, supposed to be either Shen-nung's successor or a descendant of Huang-ti.

[237] A constant refrain in the "Li sao."

[238] See note 200.

[239] 庶進德以及時： cf. *I ching* No. 1 (文言)： 君子進德脩業，欲及時也 "The superior man improves his virtue and refines his achievements, in the hope that he will be ready if the chance offers."

[240] CHANG Shih-chih 張釋之 (T. 季) served ten years without promotion until YÜAN Ang 爰盎 recommended him to the Emperor Wen; cf. *Han shu* 50.1a-b.

[241] FENG T'ang 馮唐 as Chief of Palace Secretaries 郎中署長, found occasion to protest the Emperor Wen's treatment of WEI Shang 魏尙, Prefect of Yün-chung 雲中守, who had been unjustly punished. FENG T'ang used this as an example of the Emperor's inadequate rewards for the deserving (*Han shu* 50.6a-b). T'AO Ch'ien's wording suggests that he had in mind a different version of the story.

[242] "P'ANG Kung . . . said to the King of Wei, 'If a person were now to say there is a tiger in the market place, would you believe it?' The king said no. 'If two men said so, would you believe it?' 'No.' 'Would you believe it if three men said so?' 'I would believe it.' 'It is obvious that there are no tigers in the market, but the testimony of three men creates a tiger in the market.'" (*Han Fei tzu* 9.5a). This illustrates the power of unfavorable publicity.

[243] CHIA I was Senior Tutor 太傅 to Prince Huai of Liang; cf. *Shih chi* 84.8b.

[244] For 紆遠轡 cf. No. 9 of the series "On Drinking Wine" (*Works* 3.23a) 紆轡誠可學，違己詎非迷 "It is possible to learn to hold oneself in check,/But it is really wrong to go against oneself."

I am distressed that Minister Tung's [245] profound learning
Should have endangered him repeatedly, though he
    fortunately escaped.
55 I am moved that the wise man is without a comrade—
My dripping tears wet my sleeve.

One may acknowledge the Former Kings' excellent dictum
That Heaven knows no favorites.[246]
One may find guidance by holding strictly to the One [247]
60 And by constantly aiding the good, help the cause of virtue.
But [Po-]i in his old age suffered from long hunger
And [Yen] Hui died young after living in poverty.
I lament the necessity for begging a cart to buy his coffin,[248]
I grieve the death of him who ate herbs.[249]
65 Though the one loved learning and the other practiced
    righteousness
Their lives were hard and their deaths bitter.
I suspect that this teaching [250] is no more than empty words.

It is not that in all the world there are no men of ability,
70 But it is seldom that all roads are not blocked.[251]
The men of old were burdened with care,
Worried lest they fail to make a name for themselves.[252]

---

[245] TUNG Chung-shu was "minister" 相 in the court of the Prince of Chiang-tu; cf. *Han shu* 56.21a.

[246] Cf. *Shu ching* 5/17/4: 皇天無親,惟德是輔 "Great Heaven has no affections;—it helps only the virtuous" (LEGGE, *The Chinese Classics* 3.2.490).

[247] 澄得一以作鑒: cf. *Huai-nan tzu* 16.1b (*SPTK* ed.): 人莫鑑於沫雨而鑑於澄水者 "We get a reflection, not from dripping rain, but from still waters." T'ao Ch'ien is using *chien* "mirror" in its symbolical sense; but as with water, the "one" provides a guide only if "clear," that is, not agitated; hence *ch'eng*. For 得— cf. *Tao te ching* B.2a.

[248] Cf. *Analects* 11/7: "When Yen Yüan died, Yen Lu begged the carriage of the master [to sell] and get an outer shell for his [son's] coffin" (LEGGE, *The Chinese Classics* 1.239).

[249] I. e., Po-i. I have paraphrased this line.

[250] Of the former kings; cf. line 57. This sentiment is repeated in "On Drinking Wine" No. 2 (*Works* 3.20b): 善惡苟不應、何事立空言.

[251] 罕無路之不澀: This seems to say just the opposite, and I take *han* as emphatic.

[252] Cf. SSU-MA Ch'ien, line 23.

[Li] Kuang began his career from the time he came of age,
And need not have been ashamed to be made lord of ten
    thousand households.[253]
75 But his valor was broken by a royal favorite [254]
And in the end he got not a foot of territory.
He left behind him a reputation for sincerity and integrity
To move to tears everyone [who heard of his death].[255]
[Wang] Shang offered good advice to reform corrupt
    practices;
80 He was at first listened to, but misfortune overtook him.[256]
How easily prosperous times change,
How quickly misfortune dominates.
Blue Heaven is far off,
While man's striving has no surcease.
85 Sometimes [Heaven] is responsive, sometimes it remains
    unmoved—
Who can fathom its principles?
Better endure hardship and follow one's inclinations

---

[253] Cf. Li Kuang's biography (*Han shu* 54.7a): 臣結髮而與匈奴戰 "I have been fighting the Hsiung-nu since I came of age." He was never offered such a reward, in spite of his great services. The Emperor Wen said of him, "It is too bad Kuang was born at the wrong time. If he had lived under Kao-tsu, it would not have been too much for him to have been enfeoffed as Marquis with ten thousand households" 惜廣不逢時，令當高祖世，萬戶侯，豈足道哉 (*Han shu* 54.1a). For Li Kuang's own complaint about his treatment, see *ibid.*, 54.6a. T'ao Ch'ien has 邑 "cities" for 戶, probably for the rhyme.

[254] 戚豎: i.e., Wei Ch'ing 衛青 (half-brother of one of Wu-ti's favorites) who was Commander-in-Chief of the expedition against the Hsiung-nu when Li Kuang lost his way and was late at their rendezvous. Wei Ch'ing reported him, and Li Kuang killed himself.

[255] "When he died all in the empire shed tears, whether they were acquainted with him or not. Such was his inmost sincerity and integrity with gentlemen" (*Han shu* 54.23b, after Ssu-ma Ch'ien's appreciation in *Shih chi* 109.9b).

[256] There was a false report of an impending flood in Ch'ang-an. The Emperor Yüan summoned his counselors, and Wang Feng advised the Emperor to take to a boat along with the Empress and the women of his harem, while everyone else climbed up on the city walls. Wang Shang remarked that there had never been a flood demanding such drastic precautions even in times of unprincipled government, and that under the current enlightened rule it was unlikely there was to be any flood. The Emperor was convinced, and it turned out to be nothing but a rumor. Wang Feng was embarrassed and resentful, and later secured Wang Shang's demotion (*Han shu* 82.1b-5b).

84

Than compromise and harass oneself.
Since I take no glory in the cap and carriage of office
90 Why be ashamed of tattered garments? [257]
Indeed I have missed my chance by choosing simplicity,[258]
But I shall be happy to return to the quiet life.[259]
Cherishing my feelings in solitude,[260] I shall end my years
Declining any offers [261] from the market place.

The theme of the unemployed sage, the neglected scholar, the slandered statesman is far more ubiquitous in Chinese literature than any love poetry, however chaste. History and legend provide an almost inexhaustable supply of prototypes; and legend, history and literature coalesce in the figure of CH'ü Yüan to produce the perfect representative of the type. The many Han and Six Dynasties *fu* which are dedicated to this theme are permeated by the "Li sao" to an extent hard to demonstrate in terms of verbal borrowings, though those are frequent enough. The specialized subspecies represented by the "Gentlemen Born out of their Time" is not directly modeled after the "Li sao" as the "Distant Wandering" and the "Meditation on Mystery" are, but the same lament over an unsympathetic world which affords no place for integrity or genius is the dominant motif. While this was an attitude fashionable in Later Han and Six Dynasties China, it seldom appears in T'AO Ch'ien's poetry in so obvious a form. Except in the series of seven poems "Celebrating Impoverished Gentlemen" 詠貧士, his frequent references to recluses like JUNG Ch'i-ch'i or the Four White Heads are to express admiration for the course they chose rather than to criticize directly the condi-

---

[257] 縕袍: cf. *Analects* 9/27: "The Master said, ' Dressed himself in a tattered robe quilted with hemp, yet standing by the side of men dressed in furs, and not ashamed; —ah! it is Yu who is equal to this" (LEGGE, *The Chinese Classics* 1.225).

[258] 取拙: cf. T'AO Ch'ien's poem "Returning to the Country to Live" (*Works* 2.5b): 守拙歸園田.

[259] A recurrent theme in T'ao's poetry.

[260] 擁孤襟: I take this figuratively, by analogy with 擁懷累代下 (*Works* 2.21a) and 擁勞情 ("Hsien ch'ing fu," line 77).

[261] 良價: cf. *Analects* 9/12: "Tzu-kung said, ' There is a beautiful gem here. Should I lay it up in a case and keep it? Or should I seek for a good price and sell it?' The Master said, ' Sell it! Sell it! But I would wait for one to offer the price'" (LEGGE, *The Chinese Classics* 1.221).

14

tions that made their retirement necessary. He ordinarily finds
fault with the present by eulogizing a golden age of the past, as
in the lines [262]

> I hark back to the time of Tung-hu
> When harvested grain was left in the fields overnight; [263]
> And people thumped their full bellies complacently,[264]
> Rising in the morning, returning home to sleep at night.
> Since I did not get to live in such a time
> I shall just go on watering my garden.

The themes of withdrawal from present disorder and of a golden
age of the past are neatly combined in the utopia of the " Peach
Flower Spring," an often translated anthology piece.[265]   T'AO
Ch'ien's personal interest in men out of harmony with their times
is thus well attested, but his treatment of the theme in his " Gentlemen " fu is not characteristic; it is, however, very much in line
with the pre-existing fu which he used as models. Though direct
borrowings are fewer than in the " Stilling the Passions " series,
the inspiration is strongly traditional, as even a casual reading
shows. Again I should like to make a more detailed study of T'AO
Ch'ien's poem to demonstrate how he combines conventional
themes.

Where his predecessors launched immediately into their complaint, he begins by stating his premise (lines 1-8) : man is unique
in being endowed with intelligence, and of all men the sages are
outstanding for possessing that endowment to a higher degree.
That some sages live an active life, benefiting their fellows, while
others retire and devote themselves to self-cultivation reflects a
difference in the opportunities presented them; it is the result of

---

[262] From the poem about the burning of his house, *Works* 3.17a.

[263] Ho Meng-ch'un quotes from *Tzu-ssu-tzu* 子思子: " In the time of Tung-hu
Chi-tzu, people walked straight down the road without picking up things left there,
and surplus grain was left overnight in the fields."

[264] A reference to *Chuang tzu* 4.15a: " In the time of Ho-hsü the people stayed at
home without being conscious of what they did; they went without being aware of
where they were going. They ate and were happy, drummed on their bellies and
enjoyed themselves."

[265] H. A. GILES, *Gems of Chinese Literature* 104-5.

circumstance determined by fate, and calls for neither censure by others nor complaint by the less favored ones. However, in the course of time categories of behavior are set up, subject to praise and blame. The sage regards these arbitrary standards as a snare and withdraws from the world to live in poverty and obscurity. He regrets the change of times, but finds pleasure in his enforced retreat (lines 9-18). Since the primordial state of undifferentiated being has degenerated to admit good and evil, one must choose what one's conduct will be, and naturally it is the good to which one gives allegiance, and the good is that defined in the Confucian ethic. A man who aspires to make a name for himself must keep his conduct within these limits (lines 19-28).

This introduction provides the frame of reference for the lament which begins (lines 29-36) with a bitter indictment of the world —suspicious of excellence, skeptical of integrity, slanderous of worth: the good man finds little credit for his ideals. Beginning with line 38 the complaint takes on a more personal tone, though the poet does not use the first personal pronoun here or anywhere in the poem. What he now says may apply to himself, but it is still expressed in general terms. Line 38 carries a reminder of the unattainable ideal: even in the scheme of legend which passed for early history in pre-modern China, Shen-nung and the Emperor K'uei are shadowy pre-historic figures, well buried in a past antedating Yao and Shun to whom Confucian folklore was prone to appeal. In the modern world the poet finds that virtue is no adequate qualification for a position, and illustrates the point with two examples from the Han dynasty, one of which does not seem to be very apt. The credence given false reports is introduced as a possible reason for the neglect of these men and of two other well-known Han statesmen, CHIA I and TUNG Chung-shu, whose very superiority contributed to their lack of success. Lines 55-56 bring in a motif from TUNG Chung-shu's *fu*—the wise man's isolation and need for a companion—only to abandon it without any further development.

The idea that Heaven is just, rewarding the man who devotes himself to the good, is examined in the light of precedent: from what happened to Po-i and Yen Hui it looks as though this is not a valid assumption (lines 57-68).

87

There are good men in the world, but they seldom get a chance to aid mankind as they might. This is of great concern to them, for men of ability are always anxious to put their talents to use. But even when they do have an opportunity to serve, they either end a life of achievement and devotion in disgrace—like Lɪ Kuang —or find themselves out of favor in spite of their sage counsel— like Wᴀɴɢ Shang, both of whom were undone by sycophants and slanderers (lines 69-80).

Fortune is fickle, and disaster waits on prosperity. Heaven, if not actually malevolent, is at best indifferent to human striving. If Heaven sometimes appears responsive, it is as often unmoved, and there is no way to determine on what principles it operates. If this be true, the only rational basis of human behavior is to follow one's own ideals, giving up any idea of serving a ruler, for any official career inevitably will involve compromise and trouble (lines 81-88).

If one is not impressed by the trappings of officialdom, he can be equally indifferent to the poverty which will be his lot if he refuses to seek office. The quiet life is incompatible with success, but it has its compensations, and it is better to remain untempted by specious offers which promise fame and worldly status (lines 89-94).

The argument of this *fu* is essentially the same as Ssu-ᴍᴀ Ch'ien's: virtue can look neither to the way of the world nor to the Way of Heaven for its reward. Their conclusions are similar; Ssu-ᴍᴀ Ch'ien's " entrust yourself to the spontaneous " is happily combined with T'ᴀo's " return to the quiet life " in another of T'ᴀo's poems (" The Return "), where the two injunctions are complementary. The difference in development in this *fu* owes something to Tᴜɴɢ Chung-shu, whose emphasis on companion-ship is missing in both the others, but whose elaboration in terms of historical examples was imitated by T'ᴀo Ch'ien. Of the three, Ssu-ᴍᴀ Ch'ien's best conveys the mood of bitterness and frustra-tion, but T'ᴀo Ch'ien has achieved the most subtle presentation of the dilemma confronting the man of good will, torn between his desire to serve, his dedication to ideals of conduct which require him to serve, and the unhappy state of the world where service involves the compromise of those very ideals. His choice of a life of obscurity is in part motivated by the wish to escape the disasters which overtake high-minded bureaucrats, but also it is because of his conviction that martyrdom does not further the cause of the right. The appeal to self-interest helps keep the poem above the level of banality and downright pose. Although one may prefer Ssu-ma Ch'ien's treatment of the theme, T'ao Ch'ien in this *fu* shows technical versatility and achieves a convincing statement of a complex idea.

In "Stilling the Passions" and "Lament for Gentlemen Born out of their Time" T'ao Ch'ien was writing conventional *fu* on established themes. His prefaces said as much, and an examination of his models amply confirms his statement. When he came to write "The Return" he made no such prefatory appeal to tradition, but described instead the personal experience which moved him to write: [266]

I was poor, and what I got from farming was not enough to support my family. The house was full of children, and the rice-jar was empty. I could not see any way to supply the necessities of life. Friends and relatives kept urging me to become a magistrate,[267] and I had reluctantly come to think I should do it, but there was no way for me to get such a position. At the time I happened to have business abroad [268] and made a good impression on the grandees as a conciliatory and humane sort of person. Because of my poverty an uncle [269] offered me a job in a village under his jurisdiction, but the countryside was still unquiet [270] and I trembled at the thought of going so far away from home. However, P'eng-ts'e was only thirty miles away from my native place, and the yield of the fields assigned the magistrate was sufficient to keep me in wine, so I applied for the office. Before many days had passed, I longed to give it up and go back home. Why, you may ask. Because my instinct is all for freedom, and will not brook discipline or restraint. Hunger and cold may be sharp, but this going against myself really sickens me.[271] Whenever I have been involved in official life I was mortgaging myself to my mouth and belly, and the realization of this greatly upset me.

---

[266] *Works* 5.7b-8b.

[267] 長吏 is a superior clerk or a high official. It is the former which is meant here.

[268] 會有四方之事: Li Kung-huan says this refers to the occasion when he was sent to the capital in the capacity of Secretary to the Garrison Commander 鎮衞軍建威參軍. Liu Lao-chih 劉牢之.

[269] T'ao K'uei 夔, according to T'ao Chu.

[270] 風波未靜: cf. "On Drinking Wine" No. 10: 道路迴且長、風波阻中塗 "The way is far and long, / Wind and waves (*sc.* civil disturbances) block the road" (*Works* 3.23b).

[271] For 違己 cf. note 244.

I was deeply ashamed that I had so compromised my principles, but I was still going to wait out the year,[272] after which I might pack up my clothes and slip away at night. Then my sister who had married into the CH'ENG family died in Wu-ch'ang, and my only desire was to get away. I gave up my office and left of my own accord. From mid-autumn to winter I was altogether some eighty days in office, when events made it possible for me to do what I wished. I have entitled my piece " The Return "; my preface is dated the eleventh moon of the year *i-ssu* (405).

His failure to mention any models for this *fu* does not of course mean that there were none, but it does suggest that he was not primarily concerned with imitation and elegant variation. Actually there were several *fu* extolling the bucolic life at the expense of city living, and celebrations of the seasonableness of nature in *fu* form to which he could have appealed and which may have influenced what he wrote. It is instructive to take a look at a couple of specimens: it helps explain why " The Return " enjoys a unique place in the voluminous *fu* literature while these others are seldom noticed.

It is CHANG Heng again who provides the earliest known example of a *fu* on this subject.

Returning to the Fields [273]

In the city I have spent time without end
With never a word of good counsel to benefit the
    commonweal
Fruitlessly standing by the stream and admiring the fish [274]
And waiting in vain for the River to run clear.[275]
5 I lose hope when I think of the unhappy Ts'ai Tse

---

[272] 一稔: " for one harvest," presumably for the wine which he intended to brew from the rice crop.

[273] 歸田賦. *WH* 15.25b-27a; *IWLC* 36.12b-13a; *CHHW* 53.9b-10a. I have followed the *WH* text and commentaries.

[274] Cf. YANG Hsiung's " Ho-tung fu " (*CHW* 11.6a): 雄以爲臨川羨魚不如歸而結網 " It seems to me that standing by the stream and admiring the fish is not so good as going home and tying a net." LI Shan quotes a similar sentence from *Huai-nan tzu*, but I cannot locate it in current editions of that text.

[275] I. e., for recognition. The allusion is to the " Chou poem " quoted in *Tso chuan* (Hsiang 8): 俟河之清、人壽幾何 " How long does a man live, / That he can wait for the River to run clear? "

Whose doubts were resolved by Master T'ang.[276]
Truly Heaven's operations are unfathomable: [277]
I shall emulate the Fisherman and share his joys.[278]
Overstepping the dust of the world I shall go far away
10 And take final leave of worldly affairs.[279]

Then

In the best month of mid-Spring
    When weather is fair and air clear
Highlands and lowlands burgeon
    All plants are in bloom;[280]
15 When the osprey drums his wings
    And the oriole sings his sad song,
With necks crossed they fly up and down
    Chirp-chirp, twitter-twitter.
Among such I saunter [281]
20   For the pleasure it gives me.[282]

And then

The dragon sings in the great marsh
    The tiger roars on the mountain.[283]
Above I let fly the thin silk thread,[284]

---

[276] Ts'AI Tse 蔡澤 was an itinerant politician who had been unsuccessful until he met the physiognomist T'ANG Chü 唐舉, who advised him that he had a life expectancy of another forty-three years. He subsequently became minister in Ch'in. Cf. his biography in *Shih chi* 79.15a-b. CHANG Heng implies he has no T'ANG Chü to reassure him.

[277] Cf. Ssu-MA Ch'ien's "Lament" line 15, also T'AO Ch'ien's, line 83, and note 189.

[278] Since Heaven's whims are unpredictable and I cannot expect preferment at court, I shall enjoy the irresponsibility of retirement. "The Fisherman" of course is the one who tried to reason with the intractable CH'Ü Yüan (cf. *CT* 7), though there may be also an oblique reference to the fisherman (in *Chuang tzu* 10.5b-11a) who so effectively humbled Confucius.

[279] Lines 9-10 are omitted in *IWLC*.

[280] 仲春令月 ... 百草滋榮: cf. Ts'AO Chih, "Chieh-yu *fu*" 節遊：仲春之月,百草以生 (*Works* 1.3b).

[281] 於焉逍遙: this line is from *Shih ching* 186/1 (LI Shan).

[282] Lines 16-20 are omitted in *IWLC*.

[283] Cf. *Huai-nan tzu* 3.2a: "The tiger roars and the valley wind begins to blow, the dragon rises and rain clouds gather." The poet implies that like the dragon and the tiger he is in tune with the forces of nature.

[284] Which is attached to an arrow, used in fowling. The following fishing and

Below I angle in the ever-flowing stream.
25 [The birds] collide with the arrow and die
   [The fish] covet the bait and swallow the hook.
I bring down from the clouds the bird lost from the flock
   I dangle [on my line] the *sha-liu* fish in the depths.

   Just then
The declining rays of the Great Luminary [285]
30 Are carried on by Wang-shu.[286]
So entranced by this pleasure-jaunt
I forget fatigue, though the sun is setting.
I take to heart the admonition handed down by Lao-tzu,[287]
And turn my course to my rustic hut.
35 I pluck rare melodies on the five-stringed [lute]
And recite the works of Chou[-kung] and Confucius
In high spirits I take up brush and ink and write,
To celebrate the laws of the Three Emperors.
If I set free my heart outside the realm of things
40 What are the paths of glory to me?

The affinities of this theme with the " Gentlemen Born out of their Time " are clear enough: the world of affairs is a bad place, unpredictable at best, and the wise man knows enough to get out. But the emphasis is very different. Here the " dust of the world " is quickly dismissed, and the poem describes the positive pleasures of retirement in the country. It is a Taoist theme, prominent in *Chuang tzu*, and free from the carping bitterness which permeates the *fu* of the " Gentlemen " series. The joys of country living are also described in an untitled essay by CHUNG-CH'ANG T'ung,[288] written in a strictly parallel prose that differs from the *fu* form

---

fowling motifs are common to bucolic poetry of this period; cf. the first of Hsɪ K'ang's " Poems on a Drinking Party " (*Works* 1.13a): 輕丸斃翔禽、纖綸出鱣鮪 " With light pellets we slay soaring birds, / With slender lines draw forth sturgeon."

[285] 曜靈: i. e., the sun (Lü Hsiang). Cf. Ts'ao Chih's " Chieh-yu *fu* ": 怨 | | 之無光 " I resent it that the Great Luminary gives no light " (*Works* 1.4a).

[286] 望舒 the charioteer of the moon; cf. WANG I's com. on the " Li sao," *CT* 1.19a.

[287] Cf. *Tao te ching* A/6a: " To go galloping on the hunt drives the mind to madness."

[288] 仲長統 quoted in his biography in *Hou Han shu* 79.13a-14a; *CHHW* 89.9a-b.

by its more varied rhythm and the absence of rhyme. Though not a *fu* it will serve to document this particular literary tradition:

If I might have for my dwelling
　　A spacious house and fertile fields,
　　Backed by hills and verging on a stream,
　　Surrounded by waterways and ponds,
　　Dotted with bamboo and trees,
　　Threshing floor tamped in front,[289]
　　Fruit orchard planted behind;
　　With boat and carriage to save me the trouble of walking
　　　　and wading,
　　With servants to spare me the toil of my four limbs;
　　My parents might have all delicacies for food,
　　My wife and children might lack the trials of exertion;
　　When my friends congregate I could set out wine and food
　　　　for their enjoyment,
　　And on feast days make offerings of steamed lamb and pork;
　　I would loiter in the garden
　　Or wander through the woods,
　　Splash the clear water
　　Or chase cool breezes,
　　Angle for the swimming carp
　　Or shoot at the high-flying goose;
　　Recite poetry below the altar for rain sacrifices
　　And return singing to the high hall; [290]
　　Or I would compose my mind in an inner room, meditating
　　　　on Lao-tzu's mysterious emptiness;
　　Practicing breath control, I would seek to become an Adept;
　　Or with enlightened friends I would discuss metaphysics
　　　　and books,
　　Contemplate Heaven and Earth,
　　Consider the human state;
　　I would pluck the classic melody of the Nan-feng,[291]

---

[289] 場圃築前: cf. *Shih ching* 154/7: 九月築場圃 . The *p'u* is a vegetable garden which was pounded hard for threshing in the fall.

[290] Cf. *Analects* 11/25.

[291] Shun made the five-stringed lute and used it to accompany the song " Nan-feng "; cf. *Li chi* 11.10b.

Playing a lovely tune in the clear *shang* mode;
I would take my ease above the world,
Looking with detachment on all between heaven and earth:
Untouched by the censure of my fellows,
I would live out my allotted term of life.

Then

Soaring to the heavens, I would be outside the bounds of
    the universe;
Why should I desire to have entry into the king's palace?

In this as in CHANG Heng's *fu*, bucolic pursuits are combined
with intellectual exercises, while the commitment to Taoist metaphysics is even more definite. The description of the flowering
of springtime in lines 11-20 of CHANG Heng's *fu* has no counterpart here, but it occurs prominently in a *fu* " An Excursion in the
[Spring] Season " 節遊 by YANG Hsiu 楊修 (173-219) and in a *fu*
with the same title by Ts'AO Chih (192-232). While neither of
these is an imitation of CHANG Heng, the *fu* by CHANG Hua which
borrows his title " Returning to the Fields " is, and might be
taken as an example of what T'AO Ch'ien did not write.

Returning to the Fields [292]

CHANG Hua

I obey the rhythm of *yin* and *yang*
Conforming to the seasons as they fold and unroll.
In winter my dark (?) dwelling is in the city
In spring I wander free around my country hut.
5 I go back to the old site of Chia-ju [293]
In quest of quiet [294] to live in retirement.
I cultivate plants that they may flourish
Following the hills and contours of the land.
I set out in thick clumps the vegetables and fruits,

---

[292] *CCW* 58.1b; *IWLC* 36.
[293] 郟鄏: the old capital of the Chou; cf. *Tso chuan* (Hsüan 3).
[294] 言託靜: I do not understand this phrase.

10 Raise mulberry and hemp in profusion.
   I supply my needs by taking advantage of Heaven's Way
   And amuse myself by growing herbs and drugs.
   Sometimes I wander by the banks of the Lo River
   Or perhaps stand still as it suits my fancy.
15 I eye the white sand and the piled-up pebbles
   And familiarize myself with the different flowers.
   I splash the white waves to wash my feet [295]
   And float down clear ripples as it fits my mood.
   I hesitate and stop
20 I rest amid foliage,
   My soul lodged in the infinitesimal,
   My spirit departed beyond the horizon.
   The soft grass is my mat
   The hanging shadows are my canopy.
25 I watch the high birds mount the wind
   I look down at the *t'iao* fish in the clear shallows.
   I look at the world of men, regarding it from afar,
   Cultivating spontaneity, universally valid,[296]
   That I may retire to one valley [297]
30 And long reside in obscurity, renouncing fame.[298]

" The Return " may now be read in proper perspective. The
theme is the same; actually the series of poems " Returning to
the Fields to Dwell," which T'AO Ch'ien wrote at about the same
time, use a title which could apply as well to his *fu*.

---

[295] This is rather flippant; cf. the " Ts'ang-lang Song " in *Mencius* 4A/9: " When the
water of the Ts'ang-lang is clear/It will serve to wash my cap./When the water of the
Ts'ang-lang is muddy / It will serve to wash my feet."

[296] Cf. Hsɪ K'ang's " Poems to his Elder Brother," No. 19 (*Works* 1.3b): 流俗難
悟，逐物不還，至人遠鑒，歸之自然   " Men of the world are hard to awaken,
/ They go off in pursuit of things and never return. / The Adept views such from afar
(with detachment), / And returns to the spontaneous."

[297] Like the frog in *Chuang tzu* 6.25a, who was content to be master of " the water
in one valley " 一壑之水 and stay in his abandoned well.

[298] 否 is the inauspicious hexagram No. 12; 泰, which portends prosperity, is No. 11.

95

The Return [299]

To get out of this and go back home! [299a]
My fields and garden will be overgrown with weeds—
    I must go back.[300]
It was my own doing that made my mind my body's slave [301]
Why should I go on in melancholy and lonely grief?
5 I realize that there's no remedying the past
But I know that there's hope in the future.[302]
After all I have not gone far on the wrong road [303]
And I am aware that what I do today is right, yesterday
    wrong.[304]
My boat rocks in the gentle breeze [305]
10 Flap, flap, the wind blows my gown,[306]
I ask a passer-by about the road ahead,[307]
Grudging the dimness of the light at dawn.

---

[299] *Works* 5.7b-14a; *WH* 45.27a-29b. This piece has been often translated. A representative, but by no means complete list will be found in Note 89 (with the exception of PHELPS and WILLMOTT).

[299a] 歸去來兮: I am indebted to Professor YANG Lien-sheng for pointing out the hortatory force of *lai* in this line. He calls my attention to its use in FENG Hsüan's recurrent song: 長鋏歸來乎 (*Shih chi* 75.7b) and the examples cited in P'EI Hsüeh-hai's *Ku-shu hsü-tzu chi-shih* 裴學海, 古書虛字集釋 515-6.

[300] 胡不歸 from *Shih ching* No. 36: "It's no use, it's no use, why not return?" (KARLGREN, *op. cit.*, 23).

[301] 心為形役: cf. *Huai-nan tzu* 7.4b 心者形之主也 "The heart is the master of the body."

[302] Cf. *Analects* 18/5: "As to the past, reproof is useless; but the future may still be provided against" (LEGGE, *The Chinese Classics* 1.333).

[303] 實迷路其未遠: cf. "Li sao" (*CT* 1.17b): 回朕車以復路兮及行迷之未遠 "I turn my carriage and return to the road, / Not having gone far on the wrong path."

[304] Cf. *Chuang tzu* 9.13b-14a: "When Confucius was in his sixtieth year, in that year his views changed. What he had before held to be right, he now ended by holding to be wrong; and he did not know whether the things which he now pronounced to be right were not those which he had for fifty-nine years held to be wrong" (LEGGE, *SBE* 40.144).

[305] Here begins the description of the trip home, first by water, then on foot.

[306] 風飄飄而吹衣: cf. T'ao's poem to the *Hsi-ts'ao* HU (*Works* 2.22b): 飄飄吹我衣 "Flap, flap, it blows my clothes."

[307] Cf. T'ao's poem "Detained at Kuei-lin" (*Works* 3.11b): 歸子念前途 "The home-farer is concerned about the road ahead."

Then I catch sight of my poor hut—
Filled with joy I run.[308]

15 The servant boy comes to welcome me
My little son waits at the door.

The three paths [309] are almost obliterated
But pines and chrysanthemums are still here.

Leading the children by the hand I enter my house
20   Where there is a bottle filled with wine.[310]

I draw the bottle to me and pour myself a cup; [311]

Seeing the trees in the courtyard brings joy to my face.

I lean out the south window and let my pride expand [312]

I consider how easy it is to be content with a little space.[313]

25 Every day I stroll in the garden for pleasure
There is a gate there, but it is always shut.[314]

Cane [315] in hand I walk and rest
Occasionally raising my head to gaze into the distance.

---

[308] 載欣載奔: This common *Shih ching* construction is frequently used by T'AO in his four-word poems.

[309] 三徑: an allusion to CHIANG Yü 蔣詡, an official who became a recluse rather than serve WANG Mang (*Han shu* 72.30a). LI Shan quotes a now lost work by CHAO Ch'i 趙岐, the *San-fu chüeh-lu* 三輔決錄: "CHIANG Yü . . . had a hut in a bamboo grove. He cleared three paths and sought the company of no one but Chung-yang and Chung-ts'ung. Both of them were men of principle who renounced fame and would not come out of retirement."

[310] 有酒盈樽: cf. HSI K'ang's "Verses to his Elder Brother" No. 16 (*WH* 24.12a): 旨 | | |,莫與交歡 "Fine wine fills the bottle, / But I have no one to enjoy it with" (LI Shan).

[311] 引壺觴以自酌: Cf. T'AO's poem "On Drinking Wine," No. 7 (*Works* 3.22b) 一觴雖獨進、杯盡壺自傾 "Although I am drinking alone, / When the cup is empty I tilt the bottle myself."

[312] 倚南窻以寄傲: cf. "On Drinking Wine" No. 7 (*Works* 3.22b): 嘯傲東軒下 "I whistle forth my pride beneath the east window." This suggests that the direction is not a significant detail.

[313] 容膝: lit., "enough room for the knees," an allusion to *Han-shih wai-chuan* 9/23: Master Pei-kuo's wife is arguing against his accepting an offer from the King of Ch'u: "Now though you have horses harnessed four abreast and a mounted escort, still the place you occupy is only [the room] taken up by your knees" (HIGHTOWER 311).

[314] I. e., to the outside world. Cf. T'AO's poem "To Ching-yüan" (*Works* 3.14b): 荊扉晝長閉 "My rustic gate is always shut by day."

[315] 扶老: lit., "support of the aged." Defined by Ho Meng-ch'un as being of wisteria vine or twisted bamboo.

The clouds, impersonal, rise from the peaks
30 The birds, flying wearily, know it is time to come home.
As the sun's rays grow dim and disappear from view
I walk around a lonely pine tree, stroking it.

Back home again!
May my friendships be broken off and my wanderings
　　　come to an end.[316]
35 The world and I shall have nothing more to do with one
　　　another.[317]
If I were again to go abroad, what should I seek? [318]
Here I enjoy honest conversation with my family
And take pleasure in lute and books [319] to dispel my worries.
The farmers tell me that now spring is here
40 There will be work to do in the west fields.

Sometimes I call for a covered cart [320]
Sometimes I row a lonely boat [321]

---

[316] 請息交以絕游: cf. T'AO's poem " To the Registrar KUO " (*Works* 2.17b):
息交逝閒臥 " I put an end to my contacts and go to live in retirement." Also
" Celebrating Poor Gentlemen " No. 6 (*Works* 4.11a): 翳然絕交遊 " In obscurity,
he breaks off relations with the world."

[317] 世與我而相違: cf. his " Poem to LIU Ch'eng-chih " (*Works* 2.16b): 栖栖
世中事,歲月共相疏 " The world with its ceaseless striving, / With passing time
leaves me farther behind." Also " To Ching-yüan " (*Works* 3.14b): 邈與世相絕
" I have cut off connections with the far-off world."

[318] 復駕言兮焉求: cf. Ts'AO Chih's *fu* " Excursion in the Spring Season " (*Works*
1.4a): 遂駕言而出遊; also T'AO's " After an Old Poem " No. 8 (*Works* 4.5a):
吾行欲何求.

[319] 樂琴書: cf. T'AO's poem " To the Registrar KUO " (*Works* 2.17b): 坐起弄
｜｜ " I sit up and amuse myself with lute and books." Also " When I first became
Secretary to the Garrison Commander " (*Works* 3.10b): 委懷在 ｜｜ " My taste was
for lute and books."

[320] 或命巾車: cf. *K'ung ts'ung tzu* A.31b: 巾車命駕 " I have them drive my
covered cart " (LI Shan).

[321] The cart and boat are stock fixtures in these *fu*; cf. Ts'AO Chih's " Excursion in
the Spring Season " (*Works* 1.4a): 遂降集乎輕舟 " Then we descend to assemble
in a light boat "; YANG Hsiu's *fu* of the same title (*CHHW* 51.9a-b): 御于方舟
. . . 乃升車而來反 " We ride in attached boats. . . . Then we mount carriages
and come back."

Following a deep gully [322] through the still water
Or crossing the hill on a rugged path.
45 The trees put forth luxuriant foliage [323]
The spring begins to flow in a trickle.
I admire the seasonableness of nature [324]
And am moved to think that my life will come to its close. [325]
    It is all over! [326]
50 So little time are we granted human form in the world. [327]
Let us then follow the inclinations of the heart: [328]
Where would we go that we are so agitated? [329]
I have no desire for riches
And no expectation of Heaven. [330]
55 Rather, on some fine morning to walk alone
Now planting my staff to take up a hoe, [331]
Or climbing the east hill and whistling long
Or composing verses beside the clear stream: [332]

---

[322] 尋 (var. 窮) 壑: cf. T'AO's poem " Harvest in Hsia-sun " (*Works* 3.19a):
揚檝越平湖、汎隨情壑廻 " I row across the smooth lake / And let my boat
drift as it will through the turns of a gully."

[323] 木欣欣以向榮. *Hsin-hsin* can mean " joyous," and has been so translated in
this line. The attribution of such a feeling to the processes of nature is not incompatible with T'AO's thinking, but the parallel 涓涓 of the next line makes a less
subjective reading preferable.

[324] 善萬物之得時: cf. T'AO's poem " To the *Hsi-ts'ao* HU " (*Works* 2.22b): 感
物願及時 " I am moved at the way nature strives for seasonableness."

[325] 感吾生之行休: cf. his poem " An Outing on the Hsieh Stream " (*Works*
2.7b): 吾生行歸休 " My life draws to its close."

[326] Cf. note 172.

[327] 寓形宇內復幾時: cf. T'AO's preface to his " Gentlemen " *fu*: ｜｜百年而
瞬已盡.

[328] 曷不委心任去留: cf. Hsi K'ang's *fu* " The Lute " (*WH* 18.23b): 委性命
分任去留 (LI Shan).

[329] 胡爲遑遑欲何之: cf. T'AO's " Untitled Poem " No. 8 (*Works* 4.7b): 去去
欲何之.

[330] 帝鄉 is the place where an Immortal roams when he is tired of the earth; cf.
Chuang tzu 5.7b.

[331] 或植杖而耘耔: As did the retired sage in *Analects* 18/7: 植其杖而耘.
T'AO refers to him again in his poem " In Spring, Remembering my Old Farm "
(*Works* 3.13a): 是以植杖翁,悠然不復返 " And so the old man who planted
his stick, / Will never turn again."

[332] 臨清流而賦詩: cf. Hsi K'ang's " The Lute " (*WH* 18.26a): ｜｜｜,｜新｜ (LI
Shan).

So I manage to accept my lot until the ultimate
    homecoming [333]
60 Rejoicing in Heaven's command, what is there to doubt? [334]

One clue to the difference between "The Return" and T'AO
Ch'ien's other *fu* is that in it his phraseology echoes his own
poetry much more than those *fu* which might have been his
models. Some of those echoes are listed in the notes to my translation, others are too tenuous to demonstrate easily, and a third
group I wish to discuss in more detail. Throughout T'AO Ch'ien's
poetry a number of symbols recur: the bird lost from the flock,
which represents the man who can find no place among the crowd;
the lonely pine tree, standing for constancy in adversity; chrysanthemum for longevity; music and books for aesthetic and intellectual pleasures, and also as symbols of the Confucian teachings;
a solitary cloud for detachment; and above all, wine for release.
Several of these find a place in "The Return," where their use is
so unobtrusive that a casual reading fails to discover that they
carry any unusual weight. Their symbolical force is established,
not in this one poem, but through their consistent use in the
whole corpus of T'AO's poetry.

The use of symbolism is not at all uncommon in Chinese poetry,
but most poets are content to take their symbols ready-made, so
that the device merges with that of the ubiquitous allusion. The
danger of a personal, idiosyncratic symbolism is of course that it
may obfuscate rather than clarify; of all poetic tropes it is the one
most likely to impede communication at the point where it should
be most immediate. Its advantage is that it gives the poet a
flexible tool of great power which incidentally lends coherence to
his whole poetic output. The poets, like Blake and Yeats, who
are addicted to the use of this kind of symbolism are rewarded by
having all their poetry read, not just the anthology pieces. (If
Blake's so-called Prophetic Books are not read, it is because there
is a difference between a private symbolism and a personal myth-

---

[333] 聊乘化以歸盡: cf. T'AO's poem "Lament for His Cousin Chung-te" (*Works* 2.23a); 翳然乘化去 "We leave blindly, as fate takes us."

[334] 復奚疑: cf. his poem "On Drinking Wine" No. 12 (*Works* 3.24b): 何爲復狐疑.

ology.) T'AO Ch'ien's symbolism is in large part his own creation,
except where he is consciously writing derivative poetry, and in
his other two *fu* the symbols are mostly stereotyped and appear
in much of the earlier poetry with the same value given them by
T'AO Ch'ien.

As an example of the added force gained by the symbolic use
of a term, I propose to examine the occurrences of the pine tree
in T'AO's other poems, since the word occurs twice in " The Return " (lines 17, 32). T'AO's favorite symbols tend to appear in
combination and I shall not try to isolate this one from the
others, but merely use it as a convenient focal point. No special
significance should be attached to the sequence of the following
poems; the chronology of most of them is uncertain, and I am
taking them in the order in which they occur in his collected
works.

To the Registrar KUO [335]

Warmth and moisture filled the air [336] in spring
But clear chill in this white [337] autumn season:
Dew congeals, there are no more floating mists
Heaven is high,[338] the brisk air clear.
Among the low hills peaks stand out
Now distant views are unsurpassed.
Fragrant chrysanthemums gleam in the woods
Green pines cap the hills in ranks.
I admire these forms constant in their blooming [339]
Sentinel heroes beneath the frost.
With wine cup to my lips I remember the recluse [340]

---

[335] *Works* 2.18a. My translation follows SUZUKI's (*To Emmei shikai* 175-8) except
in the last three lines.

[336] 周: lit., " everywhere."

[337] 素: white is the color associated with autumn; cf. *Po-hu-t'ung* (*SPTK* ed.)
3.12b: 其色白 .

[338] Commonly said of the autumn sky; cf. "The Nine Persuasions" (*CT* 8.2a):
悲哉秋之爲氣,天高而氣清 "Sad is the autumn season, / The sky is lofty, the
air clear."

[339] I. e., chrysanthemums and pines.

[340] 銜觴念幽人: those men of old who lived in obscure retirement. Cf. HSI
K'ang's poem "A Drinking Party" No. 1 (*Works* 1.13a): 酒中 | | |.

15

Who a thousand years ago found comfort stroking you.[341]
As I cannot escape straitened circumstances. [342]
Tranquilly [343] I shall watch out this good season.

Here pine and chrysanthemum are brought together as plants which do not yield at the first approach of autumn frost (of the chrysanthemum he says elsewhere "The flower that braves the cold blooms alone" [344]); they afford consolation to the recluse who must also live through inclement times.

On Drinking Wine, No. 4 [345]

Anxious the bird lost from the flock—
The sun sets and still it lonely flies;
Uncertain, with no fixed resting place
Through the night its cry grows sadder.
A shrill sound, as it thinks of the distant refuge—
Back and forth, always seeking.[346]
At last it reaches a solitary pine tree,
Preens its feathers after the long journey.
In the harsh wind no tree keeps its leaves:
This shelter alone will not fail.
Here the bird has refuge and resting place
Never will it leave in a thousand years.

---

[341] I follow T'ao Chu in taking "you" to mean the chrysanthemums and pines. "Found comfort" is a paraphrase of 訣 (= 快), "strengthened their resolve," this also from T'ao Chu. Suzuki understands "I admire your teachings of a thousand years ago," i.e., the recluses' teachings.

[342] 檢素: This term is not attested elsewhere. T'ao Chu and Suzuki take it to mean "Looking closely into my own heart (aspirations)." Another reading is "books and letters." Both strike me as forced, and I am emending 檢 to 儉.

[343] 厭厭: I do not see how Suzuki gets "dissatisfied" out of this. It might mean "for a brief while."

[344] "In Retirement during the Autumn Festival" (*Works* 2.4b).

[345] *Works* 3.21a.

[346] 厲響思清遠,去來何依依: I have failed to get in the *ch'ing*: "a pure, undefiled" place which is also far-off. It is probably this word which leads the commentators (e. g., Chao Ch'üan-shan 趙泉山) to a political interpretation of the piece; i. e., it is meant to chide men like Yen Yen-chih who found an impure refuge by serving the Liu Sung dynasty. However, there is a variant reading | | | | 晨,遠去何所依 "A shrill sound as it thinks of the clear dawn, / Going far away with nothing to rely on." I have paraphrased *i-i*; it is a descriptive adverb: "admiringly, with longing."

On Drinking Wine, No. 8 [347]

A green pine grows in the east garden
Hidden by a mass of vegetation;
When chill frost destroys the other plants
Its lofty branches stand out prominently.
Crowded in among other trees no one notices
Planted in isolation everyone admires.
I lift a wine pot and hang it on a cold branch
From afar I regard the tree ever and again:
Born into this dream illusion
Why should I submit to the dusty bonds?

Here the pine tree again functions as a symbol of steadfastness, but the approach is somewhat different from that of the first poem quoted above. Adversity serves to single out the individual who in pleasant times is indistinguishable from the mass. The "dream illusion" of the next to last line is perhaps the only example of specifically Buddhist vocabulary in T'ao Ch'ien, and he may have been using it simply as a current expression.

In these three poems we can see how T'ao Ch'ien persuades the reader to accept his own associations with the pine tree, elevating it to the status of a symbol. Not that his every use of the word is necessarily meant to be symbolical. It appears combined with 柏 "cyprus" in one of his imitations ("After an Old Poem" No. 4 [348]) where it has the conventional association of the term with tombs; elsewhere ("On Drinking Wine," No. 14 [349]) it is a part of the scenery, a place where friends meet to drink together. But referring back to line 18 of "The Return," the simple statement "But pines and chrysanthemums are still here" takes on the added suggestion "My refuge is here and has not failed me." Line 32 ceases to be even slightly bathetic: "I walk around a lonely pine tree, stroking it" becomes a spontaneous gesture of affection, not for vegetation indiscriminately, but for a tried friend with whose solitary state he can identify himself.

[347] *Works* 3.23a.
[348] *Ibid.*, 4.3b.
[349] *Ibid.*, 3.25a.

The home to which he has returned is presented through significant details: the eager servant who runs out to meet him, his
son who waits shyly at the door, the familiar scene in courtyard
and garden. It is given additional emotional depth with great
economy by the introduction of his established symbols: the homing birds, the detached clouds, the pine tree and chrysanthemums,
and most of all, the wine bottle, in which he has professed to have
found consolation in over half of his poems.

The brief list of country pursuits (lines 41-44) is the only part
of " The Return " which draws heavily on the conventional treatment of the subject in earlier *fu*, and the break in the regular
meter at that point makes them even more obtrusive. They may
be justified by considering them as a breathing space, a pause
before the introduction of the main theme to which the concluding lines are devoted. Without them the transition from conversations and books to the observations of nature that inspire the
formulation of his philosophy of life would perhaps be too abrupt.

The final section (lines 45-60) of " The Return " is usually
referred to the poem " Substance, Shadow, and Spirit " where the
same theme of acceptance of one's lot in the face of unavoidable
death receives its supreme statement. In both poems the point
of departure is the recurrent cycle of the seasons; this association
in lines 47-8

> I admire the seasonableness of nature
> And am moved to think that my life will come to its close

is made more explicit in the opening lines of Substance's message
to Shadow: [350]

> Heaven and Earth endure forever,
> Hills and streams never change.
> Grass and trees observe a constant rhythm:
> Frost withers and dew restores them.
> Man is said to be the most sentient being
> But he alone is not like this.
> By chance he appears in the world
> And suddenly is gone, to return no more.

---

[350] *Ibid.*, 2.1a-b.

This is the premise on which T‘AO Ch‘ien based his philosophy. The inconsistencies in his several statements of that philosophy reflect changing moods and attitudes toward his premise. Sometimes it is fear: [351]

> I constantly worry lest the Great Change take me
> Before my vital powers have declined.

Sometimes he advocates making a name that will outlast death: [352]

> That fame should end when the body dies
> Is a thought that sets my emotions on fire.
> Do good and they will love you after you are gone
> Is this not worth your every effort?

Most often he reaches for the wine bottle: [353]

> I hope you will take my advice
> When wine is offered, don't refuse.

" Spirit " offers a solution which is closest to that of " The Return ": [354]

> Too much thinking harms my life
> Simply turn yourself over to fate
> Follow the waves within the Great Change,
> Neither happy nor yet afraid.
> When you should go, then simply go
> Without any unnecessary fuss.

The bleakness of this Stoicism is replaced by joyous acceptance in " The Return." It is interesting to see how the two statements, essentially alike, take on very different emotional tones. The harshness of Spirit's solution lies in the refusal to consider any of the frivolous pleasures of life. Throughout the poem attention is uncompromisingly focused on the idea of death. By recoiling from death, both Substance and Shadow had implied that life might be desirable, but Spirit removes even this consolation.

[351] *Ibid.*, 3.16b.
[352] *Ibid.*, 2.2a.
[353] *Ibid.*, 2.1b.
[354] *Ibid.*, 2.2b.

" Neither happy nor afraid," one must face life as he meets death, " without any unnecessary fuss." The penalty for the enjoyment of life is the fear of death, and Spirit would be above joy and fear.

In " The Return " T'ao achieves a larger synthesis where there is room for present pleasures and where death has become only another manifestation of the spontaneous, the natural—that which in life is his delight; hence death too can be accepted joyfully. There are no uncertainties left, not because of indifference to what may happen, but because whatever happens to the man who sees life and death in this perspective is a source of happiness.

The exalted mood created in the last lines of " The Return " appears seldom in T'ao's poetry, and its philosophy is contradicted in poems which he certainly wrote later in life. But the inconsistency which is the bane of the philosopher is the poet's privilege. His achievement in making a conventional form the vehicle for a uniquely personal expression deserves the highest praise.

ADDITIONS AND CORRECTIONS

Page 207, line 67: *For* I suspect that this teaching [250] is no more than empty words *read* I suspect that this is the way virtue is rewarded;/I fear that this teaching [250] is no more than empty words.

SOME CHARACTERISTICS OF PARALLEL PROSE

by

JAMES R. HIGHTOWER

SOME CHARACTERISTICS OF PARALLEL PROSE

James Robert Hightower
Harvard University

The term "Parallel Prose" 駢體文 is applied to the elaborate, eu-phuistic style of writing which began to take shape in the *fu* of Han times and which culminated in the anthology pieces of the Six Dynasties and early T'ang. Since the term describes a style rather than a genre, and since there are many degrees of the ornate style, it is difficult to formulate a satisfactory definition of Parallel Prose; certainly not all parallelism in prose deserves that name, nor is parallelism the sole quality of Parallel Prose. The best approach to a definition would be a historical study of the growth of parallelism in Chinese literature, both verse and prose. In this paper I shall undertake nothing so ambitious. It is my purpose here simply to describe some of the devices common in compositions which are readily recognizable as specimens of Parallel Prose. My examples are taken from two well-known pieces, the "Proclamation on North Mountain" (abbreviated PS) by K'ung Chih-kuei (447–501)[1], and the "Preface to *New Songs from the Tower of Jade*" (abbreviated YT) by Hsü Ling (507–583)[2]. These both come from a period notorious for its almost exclusive devotion to this style of writing, though other similar examples could be found as early as Latter Han and at least as late as the T'ang dynasty.

To deserve the name, Parallel Prose must employ parallelism, and it is with the varieties of parallelism that I shall begin. In his *Bunkyō hifuron*, Kūkai distinguished twenty-nine different types of parallelism[3], though to do so he had to invent categories that are by no means mutually exclusive. In fact three general categories will take care of them all: parallelism can be Metrical, Grammatical, and Phonic.

Metrical parallelism is readily apparent to the ear and can be made obvious to the eye through the sort of typographical arrangement used

---

[1] 孔稚珪, 北山移文, text in *Wen hsüan* (abr. *WH*) 43.35b–40b (*Ssu-pu ts'ung-k'an* ed.)

[2] 徐陵, 玉臺新詠序. There are two basic texts, the one prefaced to the several editions of *Yü-t'ai hsin-yung*, and the one in Hsü Ling's collected works. Most, but not all, of the variants are recorded in Chi Jung-shu 紀容舒, *YTHY* 考異 (*Ts'ung-shu chi-ch'eng* ed.). I have used the text in 徐孝穆全集 4.1a-4a (*Ssu-pu pei-yao* ed.), with notes by Wu Chao-i 吳兆宜.

[3] 文鏡祕府論, p. 89—90 (ed. of Konishi Jinichi).

in the examples given in this paper. As in Chinese poetry, the basic structural unit of Parallel Prose is the couplet. Occasional isolated single lines do occur and are functional as paragraphing devices (see below), but there are no groups of three or five or seven lines. In a series of four parallel lines there is a tendency to vary the grammatical parallelism to avoid monotony, as PS 53–54. However, PS has a few groups of four lines where the grammatical structure remains unchanged, e.g., 95–98, 111–114, 117–120. There are no examples in these two texts of six successive lines of the same structure, nor even of six lines in succession containing the same number of beats. This means that the meter is continuously varied throughout, within the limits just stated: in each couplet a rhythmic pattern is repeated, and this pattern may be carried through one more couplet but no further.

Any given couplet may employ metrical units of three, four, five, six, or seven beats, but fours and sixes predominate. There may be a series of fours (PS 1–4), or of sixes (PS 5–8), or each line may fall into a four-four or four-six pattern (YT 20–25, 1–2)[1]. For examples of six-six, six-four and other variants (as four-seven), it is necessary to go to other texts[2]. Metrical variety is increased by the regular occurrence in the longer units (those of six or seven beats) of an "empty word" 虛字, representing a weak beat, in a position which varies between next-to-last and second-from-last (PS 5–6, 7–8; YT 22–23).

Metrical parallelism, then, gives Parallel Prose its characteristic distinction from prose which is not parallel: Parallel Prose is highly rhythmic, but the rhythms are continuously varied, and even when it uses rhyme, it is not likely to be confused with verse[3].

Grammatical parallelism is not peculiar to Parallel Prose or indeed to a literary style in Chinese[4], but no language could be more adapted to the device, and Parallel Prose consciously exploits it to the last degree. It requires that every word in the first line of a couplet be matched by a corresponding word in the second line, reinforcing the metrical repetition with a grammatical repetition of the pattern. Actually this is defining it too closely. Sometimes it is necessary to substitute the

---

[1] Notice that PS does not use this more complicated double line, while it predominates in YT.

[2] E.g., 王 勃, 騰 王 閣 序, 王 子 安 集 (*SPTK* ed.) 5.1a–3a.

[3] The relation between *fu* and Parallel Prose can be troublesome. If there is any consistent formal criterion for distinguishing them, it may be in the greater metrical irregularity of the latter. During the period when Parallel Prose exists as a style of writing used for other purposes than the *fu*, the *fu* has acquired a much greater degree of regularity than it had possessed during the Former Han.

[4] Lyly is an extreme example in English. Parallel constructions come quite naturally to any writer concerned with balanced periods and formal contrasts.

word "lexical" for "grammatical," that is, the correspondence can be purely formal.

I distinguish six types of simple grammatical parallelism. The most elementary is the repetition of a word or words in adjacent lines of a couplet:

由 余 之 所 未 窺
張 衡 之 所 曾 賦

(YT 1–2, second half-lines), where the words *chih so*, occurring in the same environment, emphasize the identical structure of the parallel lines. Repetition in Parallel Prose is usually limited to such grammatical forms as these, and is quite common. These forms are never stressed in reading, and repetition is less obtrusive than conscious variation would be.

Full words permit of more interesting manipulations, but still on a very elementary level is the variation consisting of synonyms, as 作 and 爲 in YT 5–6, 細 and 纖 (YT 13–14), 雙 and 兩 (YT 33–34), 茲 and 此 (YT 95–96). All of these could as well have been interchanged. Such feeble variation is the exception.

Only a shade more involved and certainly just as obvious are paired antonyms: 斷 and 續 (PS 71–72), 入 and 出 (PS 85–86), 未 and 曾 (YT 1–2). Simple opposites of this sort are not common; contrast is usually achieved by other means.

While pairs of words belonging to the same category are less inevitable than antonyms or interchangeable synonyms, they still carry the conviction of appropriateness. The word 手 "hand" does not inevitably call up by association 腰 "waist" (YT 14–15), but the words are suitably parallel since both are parts of the body. This is a very extensive and not easily defined type of parallelism, for categories can be of all sorts and degrees. Colors (白 and 青, PS 7–8), numbers (五 and 四, YT 7–8), cosmetics (黛 and 脂, YT 34–35), fabrics (衣 and 帳, YT 26–27)—it would be possible to subsume practically all types of parallelism under this one by inventing categories sufficiently inclusive. However, there are degrees of relationship here: a laugh and a frown (笑 and 顰, YT 26–27), a sleeve and a skirt (袖 and 裙, YT 39–40) go together more easily than a cicada and a horse (蟬 and 馬, YT 29–30), while perfume and girdle ornament (香 and 佩, YT 39–40) need some extra cement to hold them in combination. Even when the type is restricted to its more obvious members, it remains the largest reservoir on which the Parallel Prose writer can draw. It allows him to repeat his pattern closely and yet give at least the appearance of advancing his argument; it avoids the fatuity while retaining the insistence of sheer repetition: What I tell you two times is true. YT 3–4 provides the perfect example:

周王璧臺之上
漢帝金屋之中

These bonds are broken in the next type, and parallelism becomes
something else than reiteration, or an affirmative matched with a denial,
or ornamental variation. In this type the paired words, though functioning
the same way grammatically, are of different categories and hence not
readily associated together; for example, 物 "things, the world" and 霞
"mist" (PS 9–10) or 窺 "spy out" and 賦 "celebrate in verse" (YT 1–2).
Commonly this type occurs in combination with one of the more obvious
types, as though to reassure the reader that a parallel was intended,
e.g., 千金 and 萬乘 (PS 11–12), where the number words "thousand"
and "myriad" belong to the same category but "gold pieces" and "chari-
ots" do not.

Finally there is a parallelism which is only apparent, that is, where the
ostensibly parallel words do not function grammatically the same way
in the parallel phrases. This I shall call Formal Parallelism. By nature
an instance of Formal Parallelism can occur only when firmly embedded
in an otherwise impeccably parallel context. Most of this couplet (PS 7–8)
consists of the simpler types of parallelism:

度白雪以方絜
干青雲而直上

tu and *kan*, both simple verbs of motion, paired colors, snow and clouds,
grammatical connectives, temporal modifiers—but here it breaks down.
Fang, instead of meaning "just now," as it often does, must be a verb
"to compare," and so offers only a specious parallel with *chih* "straightway." This leaves *chieh* "purity" matched with *shang* "ascend." The
intrusion of such far-fetched linkages, such unlikely pairs, into a smooth
progression of identities, likes, similars, and opposites, functions much
as dissonance does in harmony. It is the astringent quality needed to
keep simple chords and resolutions from cloying.

So far I have distinguished six types: Identities, Synonyms, Antonyms,
Likes, Unlikes, and Formal Pairs. These are all types of Simple Parallelism. The examples given have been of words consisting of a single graph,
but doublets may also show Simple Parallelism, commonly where they
are proper names: Yu Yü 由余 and Chang Heng 張衡 (YT 1–2) are
parallel in only one way, that is, as the names of two men. This may be
taken as an example of Unlikes, in contrast to a more closely associated
pair of names, e.g., Ch'ao-fu 巢父 and Hsü-yu 許由 (PS 37–38), who
were contemporary recluses who both refused Yao's offer of the throne.
But this closer relation between the two is not an additional complication;
we are still dealing with Simple Parallelism.

However, there is the possibility of a further complication when the lexical unit consists of more than a single part; in addition to the semantic parallel between the two words there may be a structural parallel between the components of the words. "Clerk Han" 韓掾 and "Prince [of] Ch'en" 陳王 (YT 39–40) are parallel, not only as proper names, but also in their component parts, as surname / surname, rank / rank. The more descriptive the name the easier to find a complex parallel. "Startled Phoenix" 驚鸞 and "Flying Swallow" 飛燕 (YT 39–40) are such a pair. The parallel between components is in addition to the association between the two imperial concubines[1] who were so called.

Parallels of this sort, where one or more of the types of Simple Parallelism is combined with a parallel of meaning on another level I shall call Complex Parallelism. Complex Parallelism may be further complicated by the use of allusion to introduce a third term of relationship between the parallel binomes. "East Lu" 東魯 and "South Kuo" 南郭 (PS 29–30) are related by their parallel components and their overt meanings as two place names. But as allusions, "East Lu" must be identified with a recluse of that region named Yen Ho, while Nan-kuo turns out to be no place name at all, but the surname of another recluse, Nan-kuo Tzu-ch'i; both men are mentioned in *Chuang tzu*.

It may seem unnecessary to apply the name Compound-Complex Parallelism to cover this sort of thing, for in all my examples one of the three possible terms of relationship turns out to be fallacious, that is, it is displaced by the "true" reading which a knowledge of the allusion supplies. However, the extra meaning is always there, not so much as a trap for the unwary as an added complication, another bit of word-play. If it seems outrageous to read 西施 as "West Giver" just because she is matched with "East Neighbor" (YT 15–16), what about the same pair when the first is written 西子 "Lady West" (YT 25–26)? The fact that the descriptive term "East Neighbor" is an allusion to several celebrated east neighbors who were attractive and rather forward girls makes it appropriately parallel with Hsi-shih, the famous beauty of Yüeh.

There is a further type of Complex Parallelism which I shall call Formal Complex Parallelism. The line (YT 52–53) 九日登高 is paralleled by 萬年公主. Here *chiu-jih* and *wan nien* are not merely the descriptive terms they might well be: *chiu-jih* is not "nine days" or even "the ninth day" (in general), but the specific date of a festival (like "Twelfth Night"); *wan-nien*, though meaningful as a congratulation or a pious wish, is the appellation of a Chin princess. Hence the binomes, as they differ in grammatical function and are of incompatible meanings, are an

---

[1] This is not an ideal example, for I have been unable to identify Ching-lüan, who must, however, be such a person.

example of Formal Parallelism, while their separate components are both related as Likes. It is only because of this second relationship that the Formal Parallel appears at all; contrast the remaining words in the line: *teng kao* and *kung-chu* are held together only by rhythm; they are not felt to be grammatically parallel on any level.

Complex Parallelism applies to a pair of words consisting of more than one component (i.e., binomes) parallel in more than one way. Such a word simultaneously parallel with two other words I shall call an example of Double Parallelism. Double Parallelism occurs when a word has its first parallel in the same line; the couplet structure of Parallel Prose then demands another parallel to the same word in the following line. Double Parallelism may be either simple or complex. The Simple variety occurs in YT 9–10:

穎 川 新 市
河 間 觀 津

Ying-ch'uan parallels Hsin-shih in the same line; it also parallels Ho-chien in the second line of the couplet[1]. The next example is more interesting (YT 1–2):

凌 雲 概 日
萬 戶 千 門

"Pierce-cloud" and "Level-sun" are parallel in their component parts and also as descriptive epithets of palaces (here possibly to be read directly as names of palaces), and hence provide an example of Complex Parallelism. In the next line "myriad doors" parallels "thousand gates" as two pairs of Likes. But the two pairs in line 1 also parallel those in line 2, and though "Pierce-cloud" and "myriad doors" are not parallel in their components, they are related by sense, since "myriad doors" is also an epithet of palaces, in particular those of the imperial city; likewise "Level-sun" and "thousand gates."

The connection is even more tenuous in PS 19–20:

乍 迴 跡 以 心 染
或 先 貞 而 後 黷

Hsien chen "starting out pure" parallels *hou tu* "later becoming sullied" as simple opposites; "starting out pure" also parallels "withdrawing one's steps" (*hui chi*, i.e., "retiring") semantically and by position, though not on the level of components; likewise *hou tu* and *hsin jan*,

---

[1] As a matter of fact there is undoubtedly another term of reference involved, making this really an example of Complex Double Parallelism. But the place names are so obscurely linked with the women they no doubt were intended to recall that no satisfactory associations have been suggested for any of them; cf. the notes on the translation.

leaving *hui chi* and *hsin jan* associated by contrasting meanings and the symmetry suggested by the similar pair in the second line.

Parallelism of rhythm and parallelism of sense are the basic ingredients of Parallel Prose. Parallel sounds, Phonic Parallelism, is the third general category of embellishments found in this style of writing. Rhyme, alliteration, repetition, and tonal-pattern are the phenomena exploited. Occasional and random rhyme and alliteration are not, properly speaking, examples of parallel usage, though they may be deliberately employed by a writer as ornaments; they occur in both these compositions (e.g., YT 51 連篇 *liän p'iän*; PS 59 林欒 *liəm luân*), but I shall not discuss them further. It is convenient to treat end-rhyme as a special case, and to deal with rhyming, alliterative, and reduplicative binomes together. Tones will be taken up last.

End-rhyme is not an invariable feature of Parallel Prose, but when it occurs it is used consistently and regularly, just as it is in verse. PS is rhymed all the way through, rhymes coming at the end of the second line of each couplet. The rhyme changes eighteen times, and, with one exception (the long paragraph 65–82 has four different rhymes), the change always coincides with a change in paragraph, though not with changes in meter. There is no case of a rhyme being continued through a paragraph division, and it looks as though an important function of rhyme is to reinforce the logical divisions of the piece, as indeed it does n the *fu*. (For other paragraphing devices, see below).

The only example of a binome formed by reduplication in these materials in PS 9–10: 亭亭 paralleled by 皎皎. (The 空空 / 玄玄 of PS 45–46 are free formations rather than lexical entities). Alliterative binomes are slightly more common: PS 5–6 耿介 *kɛng kai* / 蕭灑 *sieu ṣie*; also rhyming binomes: 宛轉 *iwɒn tiwän* / 陰岑 *.iəm dz'iəm* (YT 56–57), 綢繆 *d'iəu miəu* / 紛綸 *p'iuən liuen* (PS 73–74). An alliterative binome may parallel a rhyming one: 誼囂 *χiwən χiäu* / 倥傯 *k'ung tsung* (PS 69–70). A pair of free formations similarly matched occurs in PS 43–44: 長往 *d'iang jiwang* / 不遊 *piəu iəu*. The fact that binomes of this sort are ordinarily treated as units in a symmetrical structure shows that Parallel Prose writers deliberately exploited them.

One frequently comes across categorical statements about the tonal patterns of Parallel Prose[1], that they exist that is, but I do not recall any published attempt to demonstrate precisely the nature and extent of such patterns. A couple of years ago a graduate student at Harvard University, David Farquhar, made an analysis of PS from this point of view. His first important observation was that where tonal patterns did

---

[1] I for one have made such a statement (*Topics in Chinese Literature*, p. 38), and I was making it at second hand.

occur, they were describable in terms, not of four tones, but of the two categories of tones, 平 *p'ing* and 仄 *ts'e*. These are familiar from *lü shih* prosody, *p'ing* being the "level" tones, and *ts'e* ("deflected") standing for all the rest. He further noticed two types of pattern: the sequence of tones in one line could be simply repeated in the next, e.g., PS 25–26: xxox/xxox, or, more commonly, the tones would be in inverted order in the second half of a couplet, the sort of mirror-image relationship found in *lü-shih*, e.g., PS 38–39: oox/xoo. However, out of a total of sixty-two couplets he found only five exactly parallel in either of these ways. To accord better with his data Mr. Farquhar suggested a looser definition of tonal parallelism:

Type I: One member of a parallel pair is the mirror image of its mate, with one exception.

Type II: One member of a parallel pair is identical with its mate, with one exception.

These two types, he reported, account for forty-two of the sixty-two parallel pairs in the text.

A similar study of YT shows a greater degree of absolute regularity. Out of forty-eight couplets, seventeen are perfectly symmetrical, all of them being of the mirror-image type. When the twelve additional couplets conforming to the definition of Type I are added, over half of YT is accounted for. Considering the longer and more involved line structure of YT, this compares favorably with the percentage in PS. It is apparent that while the tonal symmetry is not absolute, tonal parallelism has been deliberately exploited as a prosodic element in the composition of these pieces.

Other factors than parallelism play a role in Parallel Prose. The prominence of allusion is obvious in some of the examples cited to illustrate types of parallelism. Both PS and YT use allusion for their effects, the latter in particular depending on this device in nearly every line. It is not very illuminating to catalog the various ways in which allusion is made to function in these pieces, aside from the ways in which it reinforces and complicates grammatical parallelism, but it is interesting to see how it enters into the structure of word-plays. PS 31–32 is strictly parallel:

竊 吹 草 堂
濫 巾 北 岳

"Grass hut" and "North Mountain" offer no problem, but the *ch'ieh ch'ui* "steal a blowing" is enigmatic even when matched against *lan chin* "usurp a turban." The allusion is to the *Han Fei tzu* story of the man who got a job playing the flute in King Hsüan's ensemble of three hundred flutes, in spite of the fact that he was quite incompetent. When

the king's successor insisted on hearing the performers one at a time,
the man took to his heels. The form of the phrase *ch'ieh ch'ui*[1] owes
nothing to the language of the story and is a brilliant invention.

Given the notorious fluidity of Chinese grammar, it is not always easy
to be sure when a writer is forcing language into unprecedented molds,
but there are unmistakable signs of exuberent word-play in PS 15–18:

終 始 參 差
蒼 黃 飜 覆
淚 翟 子 之 悲
慟 朱 公 之 哭

The allusion is to *Huai-nan tzu*: "Yang Chu wept on seeing a cross-road,
because it could lead north or south; Mo Tzu cried on seeing them dye
plain silk, because it could become yellow or black." Here the allusion
is split up into four fragments, which are then set in parallel pairs inter-
related in another dimension through chiasmus, i.e., the "beginning
and end" go with Yang Chu, the "black and yellow" with Mo Tzu. The
syntax of the second couplet verges on the impossible. "Move Yang Chu's
weeping" is irregular enough, but "to tear (i.e., provide tears for) Mo Ti's
grief" is extreme even in mannered writing[2].

The basic structural element of Parallel Prose is the parallel couplet.
There is also a larger structural unit, the paragraph, which is significant
both in marking stages in the development of a theme and also in determining to some extent the form of the couplets which go to make it up.
The beginning and end of paragraphs is signaled by an unpaired line,
phrase, or word. It may introduce a new subject ("Added to all that"
加 以, YT 43/44) or a further development of an old one ("Then"
爾 乃, PS 52/53). Or a series of couplets may be framed between an
introductory word and a concluding line, as the opening lines of YT
(夫 ... 其 中 有 麗 人 焉), where the conclusion applies to everything
in between and is necessary to complete the sense of the paragraph.
Paragraphs two, three, and four of PS are similarly framed, and for
each of them the reader must hold in suspension his understanding until
the final odd line is reached. This is the common Chinese trick of syntax,
where the topic of a sentence is first stated and then followed by statement about it, carried to remarkable lengths.[3]

---

[1] The variant 偶 for *ch'ieh* is weak; it accords ill with the parallel *lan chin*. It is
simply the usual form of the allusion to the anecdote and is to be rejected.

[2] It can be matched and surpassed in the *fu*; cf. 生 貔 豹 "to live the leopard,"
i.e., "to take it alive". (Ssu-ma Hsiang-ju's "Shang-lin *fu*", *WH* 8.12a).

[3] Except for the two concluding lines, the entire piece by Chung-ch'ang T'ung
which I translated in *HJAS* 17 (1954) p. 219 is so framed. Incidentally, I should have

This organizing principle is not immediately apparent to the eye in a text printed in the traditional Chinese manner, but of course it is always accessible to the reader's ear and sense of rhythm. It becomes even more striking when reinforced by rhyme, as it is in PS. In the appended texts and translations I have used margin and indentation to show the form.

It remains to say something about the value of Parallel Prose, of these two specimens anyhow. In Chinese as in English there are many literary styles: the archaic, the poetic, the plain, the ornate. There are period styles, individual styles, and styles peculiar to specific literary genres. The essential quality of style is something easier to recognize than to isolate and describe. It has obviously to do with the ordering and choice of words, and yet the simple ordering of words to create a style in prose is of that order of subtlety and complexity which made the Wheelwright despair of transmitting the mysteries of his own craft. The furbelows and embellishments of the ornate style are more accessible, and that is what I have set out to describe in this paper. Taken together they make for a kind of writing which is about as far removed from unpremediated speech as one can get. This is only another way of saying that it is a highly artificial style, but the term prejudices the case. Any prose worthy to be called a style is the product of artifice, though its end may be to conceal rather than to flaunt art.

If the fundamental key to Parallel Prose is the couplet, it is because this is what gives it its rhythms and which makes parallelism possible in the first place. The couplet is a repetition, and the first effect of the other varieties of parallelism is to reinforce the repeated pattern. It is on this underlying pattern or series of patterns that the more subtle forms of grammatical and phonic parallelism introduce their counterpoint, a series of stresses and strains. The tensions thereby created make the reading of this sort of composition exciting as well as exhausting, an exercise in verbal polyphony. It is a style admirably adapted to the development of mood and to landscape painting in words—after all, the parallel style first developed in the *fu*. It is equally well fitted for the idle display of erudition and the construction of elaborate puzzles— whence the low repute into which it had fallen by T'ang times. It has real limitations as a medium for narration or exposition.

Of the two pieces studied here, I have no reservations about PS. It seems to me an effective and amusing burlesque done with great skill and sure taste. YT uses more subtle and complicated rhythms and is

mentioned there the excellent translation by E. Balazs which had appeared five years earlier in his "La crise sociale et la philosophie politique à la fin des Han," *TP* 39 (1949) 118–120.

dedicated to a theme which permits of any number of turns and variations; consequently the repetitions that occur seem out of place in a composition so finely wrought, and one couplet at least (48–49) is quite insipid. Still, I personally like the piece, and at any rate it serves as an excellent example of Parallel Prose in its late maturity, a little more than full-blown, perhaps, even emitting the first delicate odor of decay.

I am convinced that Parallel Prose is as untranslatable as poetry, and for the same sort of reason: its excellencies are verbal, linguistic; they do not work their magic in another medium. In appending English versions of my texts, I am only providing a practical demonstration of my claim. Still these may serve as a guide to reading in the original what is after all difficult Chinese.

PROCLAMATION ON NORTH MOUNTAIN[1]

The Spirit of Bell Mountain[2], the Divinity of Grass Hut Cloister[3], hasten through the mist on the post road to engrave this proclamation on the hillside:[4]

A man who
Incorruptible, holds himself aloof from the vulgar,
Untrammeled, avoids earthly concerns,
Vies in purity with the white snow,
Ascends straightway to the blue clouds—
   We but know of such.
   Those who
Take their stand outside things,
Shine bright beyond the mist,
Regard a treasure of gold as dust and do not covet it,
Look on the offer of a throne as a slipper to be cast off,
Who are heard blowing a phoenix flute by the bank of the Lo[5],
Who are met singing a faggot song beside the Yen-lai—[6]
   These really do exist.
   But who would expect to find those

[1] There are translations by G. Margouliès, Le *"kou wen"* chinois, p. 135, and E. von Zach, *Die Chinesische Anthologie*, p. 805.

[2] 鍾 山, northeast of Chiang-ning fu 江 寧 府, is the "North Mountain" of the title.

[3] Chou Yung, against whom the piece is directed, built a retreat on Mt. Chung which he called Grass Hut, after the 草 堂 寺 which he had seen and admired in Ssuchuan. (Li Shan).

[4] 山 庭 "mountain court", by analogy with 朝 庭.

[5] I.e., the famous immortal, Wang Tzu-ch'iao. (Li Shan).

[6] Li Shan knows of no such allusion; Lü Hsiang retails an anecdote about a recluse met under those circumstances but neglects to give his source.

Whose end belies their beginning,
Vacillating between black and yellow,
Making Mo Ti weep,
Moving Yang Chu to tears[1],
Retiring on impulse with hearts still contaminated
Starting out pure and later becoming sullied—
   What imposters they are!
   Alas!
Master Shang[2] lives no more
Mister Chung[3] is already gone
The mountain slope is deserted,
A thousand years unappreciated.
At the present time there is Chou Tzu[4]
An outstanding man among the vulgar
Cultured and a scholar
Philosopher and scribe.
   But he needs must
Imitate Yen Ho's retirement[5]
Copy Nan-kuo's meditation[6],
Occupy the Grass Hut by imposture[7]
Usurp a hermit's cap on North Mountain,
Seduce our pines and cassia trees
Cheat our clouds and valleys.
Although he assume the manner by the river side
His feelings are bound by love of rank.
   When first he came, he was going to
Outdo Ch'ao-fu
Surpass Hsü-yu[8]
Despise the philosophers
Ignore the nobility.

[1] "Yang Chu (朱公) wept on seeing a cross-road, because it could lead north or south; Mo Tzu (翟子) cried on seeing them dye plain silk, because it could become yellow or black". (*Huai-nan tzu*, *SPTK* ed.), 17.14b).

[2] Shang Tzu-p'ing 尚子平, a first century recluse; see Giles, *Biographical Dictionary* (abr. *BD*) 689.

[3] Chung-ch'ang T'ung 仲長統 (179–219), *Hou-Han shu* 79; see Balazs, op. cit.

[4] Chou Yung 周顒 (?–485) *BD* 429, *Nan-Ch'i shu* 41, whose apostasy is being rebuked in the Proclamation. He is better known as an early writer on phonology.

[5] 東魯, the recluse Yen Ho, a native of Lu, who refused a gift from the ruler. (*Chuang tzu*, *SPTK* ed., 9.21b).

[6] Nan-kuo Tzu-ch'i, who reached a state of trance through meditation (*ibid.* 1.18a).

[7] The story of the inept flute player is in *Han Fei tzu* (*SPTK* ed.) 9.9b.

[8] Ch'ao-fu and Hsü-yu both refused the empire when Yao offered it to them.

His flaming ardor stretched to the sun
His frosty resolve surpassed the autumn.
He would sigh that the hermits were gone forever
Or deplore that recluses[1] wandered no more.
He discoursed on the empty emptiness of the Buddhist sutras
He studied the murky mystery of Taoist texts.
A Wu Kuang[2] could not compare with him
A Chüan-tzu[3] was not fit to associate with him.
   But when
The belled messengers entered the valley
And the crane-summons reached his hill,
His body lept and his souls scattered
His resolve faltered and his spirit wavered.
   Then
Beside the mat his eyebrows jumped
On the floor his sleeves danced.
He burned his castalia garments and tore his lotus clothes[4]
He raised a worldy face and carried on in a vulgar manner.
Wind-driven clouds grieved as they carried their anger
Rock-rimed springs sobbed as they trickled their disappointment.
Forests and crags appeared to lack something
Grass and trees seemed to have suffered loss.
   When he came to
Tie on his brass insignia
Fasten the black ribbon,
He was foremost of the leaders of provincial towns
He was the first among the heads of a hundred villages.
He stretched his brave renown over the coastal precincts
He spread his fine repute through Chekiang,
His Taoist books discarded for good
His dharma mat long since buried.
The cries and groans from beatings invade his thoughts
A succesion of warrants and accusations pack his mind.
The Lute Song[5] is interrupted

---

[1] 王孫 (l.e., Ch'ü Yüan) is the object of the plea in the "Summons to the Hermit" (*Ch'u tz'u, SPTK* ed., 12.2a).

[2] Wu Kuang threw himself into the river when T'ang wanted to give him the throne (*Lieh hsien chuan* XV; see M. Kaltenmark, *Le Lie-sien tchouan*, Pekin, 1953).

[3] He was a Taoist Immortal; cf. *Lieh hsien chuan* XI.

[4] In imitation of Ch'ü Yüan; cf. "Li sao" (*Ch'u tz'u* 1.18a): 製芰荷以爲衣.

[5] Li Shan suggests the 琴歌 of Tung Chung-shu, now lost.

The Wine Poem[1] is unfinished.
He is constantly involved in examinations
And continually swamped by litigation.
He tries to cage Chang Ch'ang[2] and Chao Kuang-han[3] of past fame
And seeks to shelve Cho Mao[4] and Lu Kung[5] of the former records.
He hopes to succeed the worthies of the Three Capital Districts
He wants to spread his fame beyond the Governors of the Nine Provinces.
　　He has left our
High haze to reflect the light unwatched
Bright moon to rise in solitude
Dark pines to waste their shade
White clouds with no companion.
The gate by the brook is broken, no one comes back
The stone pathway is overgrown, vain to wait for him.
　　And now
The ambient breeze invades his bedcurtains
The seeping mist exhales from the rafters.
The orchid curtains are empty, at night his crane is grieved[6]
The mountain hermit is gone, mornings the apes are startled.
In the past we heard of one who cast away his cap-pin and retired to
　　　the seashore[7]
Today we see one loosen his orchids and tie on a dirty cap instead.
　　Whereupon
The Southern Peak presents us with its scorn
The Northern Range raises its laughter
All valleys strive in mockery
Every peak contends in contempt.
We regret that this vagrant has cheated us
We grieve that no one comes to condole.
　　As a result
Our woods are ashamed without end

[1] There is a 酒 賦 attributed to Tsou Yang 鄒 陽 in *Hsi-ching tsa-chi* (*SPTK* ed.) 4.4a (Li Shan); also one by Yang Hsiung. It is quite possible that no specific allusion is intended, either here or in the preceeding line.

[2] He died B.C. 48 (*BD* 21); he was a successful minor official.

[3] Chao Kuang-han was another, d. B.C. 67 (*Han shu* 76.1a–6a).

[4] Cho Mao, B.C. 53–28 (*BD* 411), was a prefect who treated the people as his children.

[5] Lu Kung, (A.D. 32–112) (*Hou-Han shu* 55) was a model administrator whose district was spared by locusts.

[6] Taoist adepts used cranes for steeds in their flights through the air.

[7] Referring to Su Kuang 疏 廣 (*Han shu* 71), who retired to the seacoast. The cap-pin is the one used to hold an official's cap on his head.

Our brooks humiliated with no reprieve.
Autumn cassia sends away the wind.
Spring wistaria refuses the moon.
We spread the word of the retirement to West Mountain[1]
We broadcast the report of the resolve of East Marsh[2].
    Now today
He is hurrying to pack in his lowly town
With drumming oars to go up to the capital.
Though he is wholly committed to the court
He still may invade our mountain fastness.
    How can we permit our
Azaleas to be insulted again
Pi-li to be shameless
Green cliffs again humiliated
Red slopes further sullied?
He would dirty with his vagrant steps our lotus paths
And soil the cleansing purity[3] of the clear ponds.
    We must
Bar our mountain windows
Close our cloud passes
Call back the light mist
Silence the noisy torrent
Cut off his approaching carriage at the valley mouth
Stop his impudent reins at the outskirts.
    Then
Massed twigs shall be filled with anger
Ranked buds shall have their souls enraged
Flying branches shall break his wheels
Drooping boughs shall sweep away his tracks.
Let us turn back the carriage of a worldly fellow
And decline on behalf of our lord a forsworn guest.

---

[1] Refers to Po-i and Shu-ch'i (*BD* 1657).

[2] Li Shan quotes the line 方將耕於東皋之陽 "I shall plow on the south slope of Tung-kao" from Juan Chi's letter to Chiang Chi, declining office (*WH* 40.38b). There is a similar phrase in P'an Yo's *fu* "Autumn Pleasures" (*WH* 13.9b): 耕 | | 之沃壤 "Plow the rich soil of Tung-kao".

[3] 洗耳, i.e., the same as the waters of the Ying, where Ch'ao-fu washed his ears to remove the taint of hearing Yao's offer.

孔稚珪：北山移文

鍾山之英　　　　　　　　　亦玄亦史
草堂之靈　　　　　　　　　　然而
馳煙驛路　　　　　　　　　學遁東魯
勒移山庭　　　　　　30　智隱南部
　夫以　　　　　　　　　　竊吹草堂
5 耿介拔俗之標　　　　　　濫巾北岳
蕭灑出塵之想　　　　　　　誘我松桂
度白雪以方絜　　　　　　　欺我雲壑
干青雲而直上　　　　35　雖假容於江皐
　吾方知之矣　　　　　　　乃纓情於好爵
　若其　　　　　　　　　　　其始至也
亭亭物表　　　　　　　　　　將欲
10 皎皎霞外　　　　　　　　排巢父
芥千金而不盼　　　　　　　拉許由
屣萬乘其如脫　　　　　　　傲百氏
聞鳳吹於洛浦　　　　40　蔑王侯
值薪歌於延瀨　　　　　　　風情張日
　固亦有焉　　　　　　　　霜氣橫秋
　豈期　　　　　　　　　或歎幽人長往
15 終始參差　　　　　　　或怨王孫不游
蒼黃翻覆　　　　　　45　談空空於釋部
淚翟子之悲　　　　　　　　覈玄玄於道流
慟朱公之哭　　　　　　　　務光何足比
乍迴跡以心梁　　　　　　　涓子不能儔
20 或先貞而後黷　　　　　　　及其
　何其謬哉　　　　　　　　鳴騶入谷
　鳴呼　　　　　　50　鶴書赴隴
尚生不存　　　　　　　　　形馳魄散
仲氏既往　　　　　　　　　志變神動
　山阿寂寥　　　　　　　　　爾乃
　千載誰賞　　　　　　　眉軒席次
25　世有周子　　　　　　袂聳筵上
　雋俗之士　　　　55　焚芰製而裂荷衣
既文既博　　　　　　　　抗塵容而走俗狀

123

風雲悽其帶憤
石泉咽而下愴
望林巒而有失
60 顧草木而如喪
　至其
紐金章
綰墨綬
跨屬城之雄
冠百里之首
65 張英風於海甸
馳妙譽於浙右
道帙長殯
法筵久埋
敲扑諠囂犯其慮
70 牒訴倥傯裝其懷
參歌既斷
洒賦無續
常綢繆於結課
每紛論於折獄
75 籠張趙於往圖
架卓魯於前籙
希蹤三輔豪
馳聲九州牧
　使我
高霞孤映
80 明月獨舉
青松落蔭
白雲誰侶
澗戶摧絕無與歸
石逕荒涼徒延佇
　至於
85 還飈入幕
寫霧出楹
意帳空兮夜鶴怨
山人去兮曉猨驚
昔聞投簪逸海岸
90 今見解蘭縛塵纓
　於是
南岳獻嘲

北隴騰笑
列壑爭譏
攢峯竦誚
95 慨遊子之我欺
悲無人以赴吊
　故其
林慚無盡
澗愧不歇
秋桂遣風
100 春蘿罷月
騁西山之逸議
馳東皋之素謁
　今又
促製下邑
浪拽上京
105 雖情投於魏闕
或假步於山扃
　豈可使
芳杜厚顏
薛荔無恥
碧嶺再辱
110 丹崖重滓
塵游躅於蕙銘
汚淥池以洗耳
　宜
扃岫幌
掩雲關
115 斂輕霧
藏鳴湍
截來轅於谷口
杜妄轡於郊端
　於是
叢條瞋膽
120 疊穎怒魄
或飛柯以折輪
乍低枝而掃跡
請迴俗士駕
為君謝逋客

PREFACE TO "NEW SONGS FROM THE TOWER OF JADE"

(This summary may be helpful in finding a way through the maze of Hsü Ling's rhetoric: In the sumptuous palaces of kings and emperors live handsome women of aristrocratic birth; others there are of humble origin, chosen for their outstanding beauty and skill as entertainers. They cause a certain amount of jealousy. Knowing how to please, they spend considerable care on their dress and makeup, and are really irresistable. But their charms are not only physical: they also have a taste for writing poetry. One or another may find time heavy on her hands— she can't always be making love, and other diversions pall. She would turn to the reading of verse for distraction, but suitable works are not easily come by. To cater to such an audience, the compiler has selected a number of poems by palace ladies or about them, which, though not serious and elevating like the Classics, are still worthy of attention. In fact, when brought out in an edition de luxe, they are just the thing for lovely idle hands and minds. If the intended readers find pleasure in them, the anthology will have served its purpose).

Cloud-piercer and Sun-leveler[1]—the like of which Yu Yü[2] never spied
    upon
A myriad doors, a thousand gates—such as Chang Heng[3] once celebrated
    in verse:
Atop the jade pavilion of the Chou king[4]
Inside the golden chamber of the Han emperor[5]
Jade trees with branches of coral
Pearl curtains with hangers of shell—[6]
    There inside are beautiful women.
    They belong to

---

[1] Ling-yün was the name of a pavilion in Lo-yang (楊龍驤, 洛陽記, quoted in *T'ai-p'ing yü-lan* 177.8b); it occurs with *kai jih* as an epithet applied to pavilions in *Chou shu*, T'ung-wen ed., 6.10b (quoted by Wu Chao-i after Ku Ch'iao 顧樵).

[2] Yu Yü was the envoy from the Jung barbarians to the court of Duke Mu of Ch'in, whose extravagance in buildings he criticized (*Mémoires historiques* 2.41, *Han-shih wai-chuan* 9/24; Wu).

[3] Chang Heng's "*Fu* on the Western Capital" (*WH* 2.11a) has the phrase 門千戶萬. (Wu).

[4] King Mu of Chou built a pavilion for his favorite which was called the Double Jade-disk Pavilion 重璧臺 (*Mu-t'ien-tzu chuan*, *SPTK* ed., 6.29b; Wu).

[5] Emperor Wu of the Han, while still only a child, said that if he could have the Princess A-ch'iao, he would build a room of gold to keep her in. (*Han-Wu ku-shih*, Lu Hsün's 古小說鉤沈 ed., 337; Wu).

[6] When the Emperor Wu built a residence for the spirits (神), he planted jade trees with coral branches in the front courtyard. Inside were curtains made of white pearls with tortoise-shell hangers. (*Ibid.*, 347; Wu).

The aristocracy of the Five Tombs[1], chosen for the Side Palaces[2]
The good families of the Four Clans[3], famous in the women's quarters.
　　Besides them there are those from
Ying-ch'uan and Newmarket[4]
Ho-chien[5] and Watchford[6],
Described as Charming Beauty[7]
Surnamed Artful Smile[8].
Palace women of the Ch'u king—everyone marked the slender waists[9]
Beauties of the Wei state—all voices exclaimed at the delicate hands.[10]
Read in the Odes, versed in etiquette—not like the East Neighbor[11], who
　　was forward
Graceful and seductive, different from Hsi-shih, who had to be taught[12].

---

[1] The tombs of the first five Han emperors, near Changan, in the vicinity of which the aristocracy and well-to-do lived. (Wu).

[2] *I-t'ing* designates the apartments where the palace women were lodged; the earlier term was *yung-hsiang*, which parallels it in the next line. (*Han shu* 19A.9b) For the selection of girls of good family for palace duty, see *Hou-Han shu* 10A.3a (Wu).

[3] In the time of the Han Emperor Ming, the Fan, Kuo, Yin, and Ma families of Imperial Consorts were known as the Lesser Nobility of the Four Clans. (Ku Ch'iao).

[4] Ying-ch'uan was the native place of Empress Yü, noted for her beauty and decorum. Hsin-shih is unexplained. (Wu).

[5] Wu Chao-i refers Ho-chien to the Lady Chao of the Kou-i Palace, consort of the Emperor Wu and mother of the Emperor Chao, from whose clenched fist the Emperor Wu retrieved a jade hook (*Han-wu ku-shih* 353). She was first located in Ho-chien through a violet emanation which rose up from the ground there. See also *Han shu* 97A.16a.

[6] Kuan-chin is where the Empress Tou was born and buried (*Han shu* 97A.7b) (Wu).

[7] Chi Jung-shu (*YTHY k'ao-i*) suggests emending the unexplained 嬌娥 to 婕娥 the term for one of the fourteen grades of imperial concubines *Han shu* 97A.1b).

[8] Tuan Ch'iao-hsiao 段巧笑 was one of the Wei Emperor Wen's favorite palace ladies. She is credited with first using rouge 紫粉 on her face (Ts'ui Pao 崔豹, *Ku-chin chu* 古今注, Commercial Press 1956 ed., p. 26).

[9] The king of Ch'u prefered slender girls and as a result many of his palace women starved themselves. It is quoted as a proverb in *Hou-Han shu* 54.16b, and adapted from the *Mo tzu* passage (15) where it is applied to courtiers overanxious to please.

[10] Wu Chao-i refers this to the *Classic of Songs*, No. 107/1: 摻摻女手，可以 絳裳 "Delicate are her hands, They can sew a skirt". Chi Jung-shu also quotes No. 57: 手如柔荑 "Hands like tender shoots". A more specific reference seems called for.

[11] Ever since "The Lechery of Master Teng-t'u", (*WH* 19.12b), girls living next door on the east side have had a bad reputation, in literature at least.

[12] King Kou-chien of Yüeh had Hsi-shih trained three years before presenting her to the King of Wu (*Wu-yüeh ch'un-ch'iu, SPTK* ed., B. 39a).

Her brother was court musician: from childhood she studied singing[1]
As a girl she grew up in Ho-yang; from the first she was good at dancing[2].
Her guitar tune needs no Shih Ch'ung[3]
Her flute medley requires no Ts'ao Chih[4].
She studied lute playing in the Yang clan[5]
She learned flute blowing from the Ch'in girl[6].

To such an extent that
The news of a new favorite reached the Palace of Eternal Joy—
The empress Ch'en learned of it and was uneasy[7].
The painting showed Heaven's own Immortal—
The Barbarian Queen looked and was jealous from afar[8].

[1] Li Yen-nien's office was "Harmonizer of the Musical Pipes" 協律都尉. His sister, the Palace Dame Li, was until her death the Emperor Wu's favorite. (*Han shu* 97 A.13 a)

[2] The prefered reading in *Han shu* 97 B.10 b is 陽阿 for the name of the place where the dancing girl Chao Fei-yen was discovered by the Emperor Ch'eng on one of his incognito expeditions.

[3] The song of Wang Chao-chün, when she was sent away to the barbarians, for which Shih Ch'ung wrote a preface (琵琶引序, *Ch'üan Chin wen* 33.12 b), in which he attributed the original to the Wu-sun Princess. *Yüeh-fu tsa-lu* 29 also mentions the Wu-sun Princess as the first to play the Guitar Song. Shih Ch'ung had a singing girl named Lu-chu whom he taught to perform this song (樂府古題要解, *Li-tai shih-hua hsü-pien* v. 1, A.9 a).

[4] "K'ung-hou yin", according to *Ku chin chu* 12, is the name of the tune played on the *k'ung-hou* (a stringed instrument) by Li-yü, wife of the Korean 霍里子高, who had himself heard it sung impromptu by the anonymous wife of an old man who drowned crossing the river, the song being called 公無度河 "Don't Cross the River, Sir," from the words of the first line. It is true that among Ts'ao Chih's *yüeh-fu* poems is one entitled *k'ung-hou yin*, but it is only a conventional feasting and congratulatory song, with no echo of the lugubrious story told about the original version. The term 雜引 "medley" does not occur in any of the pre-Sung texts on *yüeh-fu* song titles; it may perhaps refer to another song of Ts'ao Chih's now lost.

[5] Yang Yün 楊惲, describing his life in retirement in his letter to Sun Hui-tsung (*Han shu* 66.11 b), says that his wife, a native of Chao, played on the *se*, presumably the Ch'in songs he knows. (Wu).

[6] Nung-yü, daughter of Duke Mu of Ch'in, fell in love with a flute-player, Hsiao-shih, who taught her to make the song of the phoenix on the flute; see *Lieh hsien chuan* XXXV. (Wu).

[7] The Empress Ch'en was enraged when she heard of the Emperor Wu's involvement with Wei Tzu-fu (who later became the Empress Wei). (*Han shu* 97 A.11 a; Wu).

[8] When the Han Emperor Kao-tsu was besieged in P'ing-ch'eng, his strategist Ch'en P'ing (*Shih chi* 65) reported to the *Yen-shih* (the consort of the Hsiung-nu chief) that the Chinese were sending a beautiful woman as a present to the *Shan-yü*, whose affection she would surely gain. The *Yen-shih's* jealousy was roused and she saw to it that the siege was raised. This is one of Ch'en P'ing's (probably quite fictitious) "secret plans"; see H. H. Dubs, *History of the Former Han Dynasty* 1.116–7 note 2.

Or further,

The East Neighbor with artful smile
   going to serve in his couch when he changed clothes[1]
West Lady with her slight frown
   about to lie across the first-class bed,[2]
Entertaining the Emperor in Sa-so[3]
   she twists her slender waist to the Tied Wind [measure][4]
At a party in Yüan-yang[5]
   she sings a new song to the once-played tune.[6]
She shapes her hair into singing cicada diaphanous locks
She mirrors falling-off-a-horse hanging coils[7]
   Pinned back with golden hairpins,
   Drawn across with jeweled comb.[8]
Stone blacking from Southtown
   draws perfectly paired moths[9]
Swallow rouge from Northland[10]
   marks accurately two dimples.

[1] Presumably Wei Tzu-fu again, whom the Emperor Wu noticed among the singers at the establishment of the Lord of P'ing-yang, to the neglect of the beauties assembled for his inspection. When he excused himself to change his clothes, Tzu-fu waited on him and gained his favor in a carriage. The Emperor returned to the party in fine spirits and presented his host with a thousand *chin* of gold. Subsequently Tzu-fu was sent to the palace. (*Han shu* 97 A.11 b; Wu).

[2] *Chuang tzu* 5.43 a mentions Hsi-shih's knitted brows (she suffered from heart trouble), which served only to enhance her beauty. The "first-class bed" described in *Han-Wu ku-shih* 347 was for the use of spirits, while the Emperor Wu himself occupied the second-class 乙 one. The 甲乙帳 mentioned in *Han shu* 65.14 a and 96 B. 23 a are merely imperial luxuries, and the meaning in the Preface is "emperor's bed".

[3] Sa-so was the name of a palace building within the precints of the 建章宮, so-called because it would take a fast horse a day to circumambulate it. (*San-fu huang-t'u*, 三輔皇圖 *Ku-chin i-shih* ed., 2.5 a; Wu).

[4] *Chieh-feng* appears three times in *Wen hsüan* (8.12 b, 14.16 a, 34.7 b) as the name of a piece of dance music.

[5] The name of the palace where the Han Emperor Ch'eng first heard of Chao Fei-yen's sister. (*Fei-yen wai-chuan, Han-Wei ts'ung-shu* ed., 2 b; Wu).

[6] *Tu-ch'ü* should be the name of a tune, to match *chieh-feng* above, but I have been unable so to identify it.

[7] Both are fanciful descriptive names of types of hairdo.

[8] 樹 must be for 梳.

[9] The standard epithet for delicately traced eyebrows is 蛾, metonymy for "moth antennae". *Shih tai* "mineral blacking" or 石墨 "graphite", is a product of the south. Nan-tu is probably not to be taken as a specific place name.

[10] *Yen-chih* is so called because it comes from the state of Yen in the north (*Chung-hua ku-chin-chu* 32); it has of course nothing to do with "swallows".

And also

The Immortal boys on the mountain

   share the pills with the Wei Emperor[1]

The precious phoenix in Yao

   presents the calendar to Hsüan-yüan[2].

The gold star[3] vies with Wedded Woman[4] for brightness

The musk moon[5] contests with Ever-Fair[6] for brilliance.

From Startled Phoenix's annointed sleeves

   on occasion wafted the perfume of Clerk Han[7]

[1] This and the following line seem quite out of place. Chi Jung-shu suggests that something has been omitted here; it would be easier to assume that the boys are intruders. In any case, I do not know what the connection between them and the ladies' dress and makeup is supposed to be. The allusion is to Ts'ao P'ei's poem (*Ch'üan San-kuo shih* 1.8a) which begins:

> How very high is West Mountain
> High, high, nearly no end!
> On top two Immortal Boys
> Know no hunger and eat no food.
> They gave me some round pills
> Shining with all the five colors.
> For two or three days I swallowed them
> Until my body sprouted wings.

[2] This line yields no sense, neither as a parallel to the preceding one nor as a part of the Preface as a whole. Wu Chao-i detects an echo of Ling Lun's establishing the musical modes for the Yellow Emperor: the song of the phoenix (male and female) gave him the notes of the *Huang-chung* mode (*Han shu* 21A.4a; *Lü-shih ch'un-ch'iu*, SPTK ed., 5.8b; *Shuo yüan*, SPTK ed., 19.21b). He quotes Ku Ch'iao, who cites Chang Yen's commentary on *Han shu* 19A.1b, elaborating on the mythical Emperor Shao-hao's use of bird-names for his officials, to the effect that a person named Phoenix-bird was in charge of the calendar; further, the Yellow Emperor (alias Hsüan-yüan) devised the cyclical terms designating years and days. All of this adds up to nothing but the certainty that the key to this line lies elsewhere. The phrase "precious phoenix at her waist" 腰中寶鳳 promises well as a development of the ornamentation theme, but such a reading cannot be reconciled with the second half line. Incidentally, there is no justification for taking *Yao* as a place name, except to balance the *ling shang* above.

[3] *Chin-hsing* is either an ornament or a kind of makeup. Wu Chao-i quotes Ku Yeh-wang 顧野王's poem (*Ch'üan Chin shih* 4.2a): "When her toilette is done the gold star shows" 收罷金星出.

[4] *Wu-nü* is the name of constellation.

[5] *She-yüeh* is a beauty mark. Wu Chao-i cites *Yu-yang tsa-tsu*: "Recently dimples in the form of a sickle moon 射月 have been the rage".

[6] *Ch'ang-o* is the moon goddess.

[7] The name of Chia Ch'ung's daughter may have been Ching-luan, though it is nowhere recorded. Her lover, Clerk Han, supplied her with tribute perfume and so betrayed their liason (*Shih-shuo hsin-yü* 3B.48a–b). It is likely that Startled Phoenix is another imperial concubine like Flying Swallow (see p. 82, note 1,) but I cannot so identify her.

On Flying Swallow's long skirt
  it is meet to tie the pendant of the Prince of Ch'en[1].
Though her portrait was never painted
  she might enter the Sweet Springs Palace all the same[2]
While you may say she is no goddess
  she romps under Sun Tower just as well[3].
  Truly it may be said of her that she is
State's Bane and City's Fall[4]
Unmatched and unrivaled.
  Added to all that
Her heaven-given sensibility is receptive and bright
Her rare mind is sharp and artistic.
Admirably schooled in the art of letters
She excells in verse and ode.
Inkstone case of crystal is daylong by her side
Brush-holder of lapus lazuli never leaves her hand.
Fresh verse fills her work basket—
  not only celebrating peony flowers[5]
New creations, page after page—
  hardly limited to the grape vine[6].
On the ninth day she climbs to a height,
  when she writes with true feeling[7].

[1] Ts'ao Chih, Prince Ssu of Ch'en, offered his girdle ornament to the Goddess of the Lo River (see *WH* 19.17b, Wu). The association with Flying Swallow is anachronistic, and the couplet should rather be read as applying to a woman lovely as Startled Phoenix or Flying Swallow, whose sleeves carry a perfume rare as that stolen by Clerk Han, and who is as suitable an object of Ts'ao Chih's attentions as was the goddess.

[2] The Emperor Wu had the Palace Lady Li's portrait placed in the Kan-ch'üan Palace after her death. (*Han shu* 97A.13a; Wu).

[3] The Goddess is the one who appeared to the King of Ch'u on Witches' Mountain saying, "At dawn I am the morning clouds, evenings the driving rain, every morning, every evening, below the Sun Tower". ("Fu of Mr. Kao-t'ang", *WH* 19.1b; Wu).

[4] Referring to the epithets used by Li Yen-nien in describing his sister to the Emperor Wu (*Han shu* 97A.13a).

[5] Wu Chao-i quotes a "Eulogy of the Peony" by Hsin Hsiao 辛 蕭, wife of Fu T'ung 傅 統: "Bright is the peony, planted in the front court, mornings moistened by the sweet dew, which dries under the midday sun's rays". It concludes "The poet of old (i.e., *Shih ching* No. 95) offered this glorious blossom, taking it as a symbol of his love; he wet his brush and wrote a song". (*Ch'üan Chin wen* 144.2a–b).

[6] The grape vine remains a mystery.

[7] This refers both to the custom of climbing hills on the festival of the Double Ninth and to the enigmatic remark in the "Essay on Bibliography" *Han shu* 30.36b) that "he who, on climbing to a height, can compose (?recite) verse will make a Great Officer".

For the Myriad Years Princess
  there is no lack of a eulogy praising her virtue[1].
  Her charm and beauty were earlier described; this tells of her talent
  and sensibility.
  On the other hand she may be
Langorous in Pepper Chamber[2]
Secluded in Mulberry Quarters[3]
Crimson crane solemn at dawn[4]
Bronze knocker quiet at noon[5].
The turn of the Triad not yet come,
  she need not serve with bedding in arms[6].
While the fifth day is still distant,
  who will practice a song?[7]
Idling with little to do
Solitary with much leisure

[1] The Wan-nien Princess was the daughter of the Chin Emperor Wu. On her death the Emperor ordered the Imperial Concubine Tso Fen 左芬, noted for her literary skill, to write the dirge. (*Chih shu* 31.11b; Wu).

[2] An empress is called *chiao fang*, from the use of aromatic pepper in the plaster on the walls of her chambers—both for fragrance and to repel noxious vapors. (*Han kuan i* 漢官儀, *TSCC* ed., B. 37; Wu).

[3] The lying-in chambers for palace ladies were so-called. (*Han shu* 97B.8a, com. by Su Lin and Chin Shao).

[4] The key which opens the palace gate in the morning is called a "crane key" 鶴籥 (see Wu Chao-i's quotation); I do not know why it should be crimson.

[5] For 蠡 read 鋪 with *Wen-yüan ying-hua* (Chi Jung-shu); it is the boss holding a ring by which the door is pulled to. It also serves as a knocker.

[6] The reference is clearly to *Shih ching* No. 21, with the concubines paying their surreptitious visits to the King, bedding in arms (抱衾與裯). There is perhaps also a reference to Shih ching No. 118, with its repeated 三星在天 (隅, 戶). Since the appearance of the Triad (a constellation) is there connected with a reunion of husband and wife, the phrase 三星未夕 could mean "This is not the evening for the meeting symbolized by the Triad". If we stick to No. 21, identifying the Triad with the 三五在東 of that poem and taking it as synonymous with the 嘒彼小星 of the first line, we get something like the translation above: that is, the lesser concubines do not spend the night with the king, their turn not having yet come.

[7] Li Shan, in his com. on the lines "Dressed in garments of thin silk, She sits at the door rehearsing the clear song" 理情曲 (*WH* 29.6a), quotes Ju Shun's com. on *Han shu*: "Today musicians speak of practicing a piece of music once every five days as 理樂". This yields a ready reading for the line of the Preface, but hardly a satisfactory one. Read in terms of the *Shih ching* allusion (itself an added element of parallelism!), it means something much more appropriate in context. *Shih ching* No. 226 has the lines 予髮曲局, 薄言歸沐 "My hair is all twisted and tangled; I shall go home and wash it". Also: 五日爲期, 六日不詹 "Five days was the appointed time, On the sixth he has not come". The "filthy and absurd view of Maou" to which Legge could only refer (*Chinese Classics* 4.412

Tired of the dilatory bell from Eternal Joy Palace[1]
Weary of the slow hours in Central Hall[2]
The slight frame has no strength—
    fearful of fulling clothes in Nan-yang[3]
A whole life spent in a secluded palace,
    she laughs at the brocade woven in Fu-feng[4].
    Even if, like
The Jade Girl, she plays tosspot,
    all pleasure is exhausted in a hundred pitches[5]
The Lady of Ch'i, she contests at *po*,
    her enjoyment is worn out with the six sticks[6].
She takes no pleasure in wasted time
But bends her mind to new verse,
    which can
Serve as heart's ease[7], subtly softening melancholy grief.
    But
Famous works of past time

note), must have caught Hsü Ling's more robust fancy. Mao remarks simply that married women must have sexual intercourse once every five days (*Mao-shih chu-su, SPPY* ed., 15.2.4b); he is only quoting a passage from the *Li chi* (12/4) defining the conjugal rights of women (under the age of fifty). Hsü Ling's line then reads "Until the fifth day comes, who is going to comb the kinks out of her hair ?"

[1] The Ch'ang-lo Palace is where the ruler's mother was lodged. The bells announce the passage of time, which must seem slow to the palace ladies who wait on her.

[2] The Empress's quarters are referred to as 中宮 in *Han chiu i*, 漢舊儀 (*TSCC* ed.) B. 12.

[3] This should be a specific reference, but I cannot identify the woman of Nan-yang who prepared winter clothes for her husband away on the frontier (the usual situation associated with *tao-i*).

[4] Fu-feng is a place on the site of Hsien-yang, the old Ch'in capital, hence Ch'in-chou, of which Tou T'ao 竇滔 was prefect. When he was transferred to the border, his wife née Su, who was skilled at brocade work, wove a palindrome into a piece of brocade, which she sent to him. (Wu)

[5] According to *Shen i ching* (*Han-Wei ts'ung-shu* ed.) 1.1a, when Tung-wang-kung and Jade Girl play tosspot, every time a throw fails to connect, Heaven laughs at them. The light flashing from Heaven's opened mouth is the lightning. (Wu) For 驍 as "a throw", see L.S. Yang, "An Additional Note on the Ancient Game *liu-po*", *HJAS* 15 (1952) 134.

[6] For 著 in the game *liu po*, see *ibid*. The allusion to the "Lady of Ch'i" is unidentified.

[7] *Hsüan* and *su* are two kinds of plant reputed to be effective anodynes, as in Wang Lang's 王朗 (d. 228) "Letter to the Heir Apparent of Wei" (*Ch'üan San-kuo wen* 22.10b): "I suggest reading as a sheer delight and as support against hunger and thirst; not even the *hsüan* plant for forgetting worry or *kao-su* for relieving fatigue are any better". (Wu).

Skilled creations of the present day
Distributed in the Unicorn Hall[1]
Dispersed in the Vast City[2]
Unless collected in a volume
Leave her no way to read them.
   This being so
Burning tallow to copy at night
Grinding ink to write at dawn,
   One has here recorded love songs enough to make ten scrolls. They
   are not fit to put alongside the Odes and hymns[3], nor are they the
   overflow from the Bards[4]; it is rather like the waters of the Ching and
   the Wei, (which flow in the same channel without commingling)[5].
   Then they are
Laid out in a golden casket
Mounted on costly scrolls
In the finest tradition of the Three Chancelleries—[6]
   the calligraphy of uncoiling dragon and twisting caterpillar.
Five colored patterned stationery—[7]
   paper of Hopei and Chiaotung[8].
In the high chamber with red powder[9]
   still establishing questionable readings[10],
Fresh incense to expel the noxious
   guards against Yü-ling bookworms[11].
   Like

---

[1] Unicorn Hall 麒麟閣 in Wei-yang Palace, used as a repository for archives during the Han. (*San-fu huang-t'u* 6.47; Wu).

[2] The name of a gate in Loyang where books were stored. (Wu).

[3] I.e., of the *Classic of Songs*.

[4] Who wrote the "Feng" poems of the *Classic*.

[5] The waters of the Ching and the Wei flow together for 300 *li* without clear and muddy intermingling (*San Ch'in chi* 三秦記, *Lung-hsi ching-she ts'ung-shu* ed., 3a). That is, this lesser poetry coexists with that classical tradition without contaminating it.

[6] For the *san t'ai* see *Han kuan i* A.21 (Wu).

[7] "Five colored paper" is mentioned in *Yeh chung chi* 鄴中記, *TSCC* ed., 1 (Wu).

[8] I can find no allusion connecting paper and these two places.

[9] "Red powder" (var. 鉛紛) is not only used as a cosmetic; for its use as the ingredient of red ink in collating and annotating texts, see E.H. Schafer, "The Early History of Lead Pigments and Cosmetics in Chinese", *TP* 44 (1956) 437–8.

[10] *Lu-yü* is a generic term for words easily confused in copying. It occurs as part of a proverb quoted in *Pao-p'u tzu* (*SPTK* ed.) 19.7a: "After three copies, 魚 becomes 魯 and 虛 becomes 虎."

[11] *Mu-t'ien-tzu chuan* 5.26a mentions "books eaten by worms in Yü-ling". (Wu).

The *Divine Flight* and the *Six Scales*[1] it is stored high in a jade box
The *Vast Radiance* and the *Immortal's Receipt* it is stuffed in the cinnabar
    pillow[2].
    There
Inside the Blue Ox Curtain[3] the old tune is not yet finished
Before the Red Bird Window[4] her fresh makeup is done.
    This is the time when she
Opens this green scroll
Unties these binding cords,
Always toying with them behind the book curtain
Forever unrolling them in slender hands.
    Not of course like
Empress Teng studying the *Spring and Autumn*—[5]
    not easy to practice the scholar's task
Empress Tou wrapped up in Huang-Lao—[6]
    unachieved the technique of gold and cinnabar[7].
    But better than
The rich man of West Shu who put all his passion in the "Lu Palace *fu*"[8]

---

[1] Two esoteric works which Hsi-wang-mu gave to the Emperor Wu; he kept them in a golden casket inside a box of white jade. (*Han-Wu nei-chuan, Han-Wei ts'ung-shu* ed., 15b; Wu).

[2] *Hung-lieh* is an alternative title for the *Huai-nan tzu. Hsien-fang* is perhaps to be taken generically as "receipts for becoming an Immortal", referring to the esoteric works found in a pillow (i.e., a headrest) by Liu Te at the time of the trial of the Prince of Huai-nan. These were the books that got Liu Te's son, Liu Hsiang, in trouble when he tried to make alchemical gold with their help. (*Po-wu chih*, 7.41–2, Wu).

[3] The blue (or black) ox is always connected with a recluse. This is the only occurrence of the phrase *ch'ing-niu chang* in *P'ei-wen yun-fu*, and I have no idea what it means.

[4] The window through which Tung-fang So peeped when Hsi-wang-mu was presenting the peaches of immortality to the Emperor Wu (*Po-wu chih* 3.17, Wu). What this has to do with the "fresh makeup" I do not know.

[5] Since *Teng* parallels *Tou*, which must be the Empress Tou (see note 6), it is likely that the reference is to an Empress Teng. I have not found one who fits exactly. Wu Chao-i suggests the Empress Ma, whose familarity with the *Ch'un-ch'iu* is specifically mentioned (*Hou-Han shu* 10A.10b). The Empress Teng (Ho-hsi 和 熹) had an education in the Classics (*Hou-han shu* 10A.14a) and was known as a patron of Confucian studies (*ibid.* 109A.2b, 110A.15a); it is possible she is the one meant.

[6] The Empress Tou, mother of the Emperor Ching, is famous for her preference for Taoism at the expense of the Confucians. (*Han shu* 88.17a, 20a).

[7] I.e., of making gold from cinnabar.

[8] Liu Yen 劉 琰 was a general of the state of Shu who had extravagant tastes. He taught "several tens" of his slave girls to recite the "Lu-ling-kuang Palace *fu*" of Wang Yen-shou (*WH* 11.17a–29b); see *San-kuo chih, Shu chih* 10.9b. (Wu).

Or the Heir Apparent in the First Lodge, always having them recite the
    "Hollow Flute."[1]
Lovely those Ch'i girls[2],
    it shall help them pass the time.
How fine the red brush—[3]
    none may criticize it[4].

徐陵：玉臺新詠序

夫凌雲概日	由余之所未窺
萬戶千門	張衡之所曾賦
周王璧臺之上	
漢帝金屋之中	

5   玉樹以珊瑚作枝
             珠簾以玳瑁爲柙

                          其中有麗人焉

其人也
| 五陵豪族 | 充選掖庭 |
| 四姓良家 | 馳名永巷 |

亦有
    潁川新市
10   河間觀津
    本號嬌娥
    曾名巧笑
| 楚王宮裏 | 無不推其細腰 |
| 衛國佳人 | 俱言訝其纖手 |

[1] While still Heir Apparant, the Han Emperor Yüan was so pleased with Wang Pao's *fu* "The Hollow Flute" (*WH* 17.15a–23a) that he had his palace attendants recite it. (*Han shu* 64B. Wu).

[2] The line is taken from *Shih ching* No. 39/1: "Lovely are those ladies of the Ch'i clan, I shall make my plans with them".

[3] Referring to *Shih ching* No. 42/2: "She gave me a red tube", where Mao identifies it as the red brush-tube used by the Female Recorder 女史, whose duty it was to keep a record of transgressions in the harem and also to keep track of the proper sequence of visits by its members to the ruler's bedchamber (*Mao-shih chu-su* 2.3. 7b–8b). Hsü Ling is delicately suggesting that the poems in his anthology are on erotic subjects, but still are above reproach.

[4] The reading for this line 麗以香奩 "Pretty the perfumed compact", as Chi Jung-shu remarks, gives no satisfactory sense.

15　　閱詩敦禮　　豈東鄰之自媒
　　　婉約風流　　異西施之被教
　　　弟兄協律　　生小學歌
　　　少長河陽　　由來能舞
　　　琵琶新曲　　無待石崇
20　　笙箊雜引　　非關曹植
　　　傳鼓瑟於楊家
　　　得吹簫於秦女

　　至若
　　　寵聞長樂　　陳后知而不平
　　　畫出天仙　　關氏覽而遙妒
　　至如
　　　東鄰巧笑　　來侍寢於更衣
　　　西子微矉　　得橫陳於甲帳
　　　陪遊馺娑　　騁纖腰於結風
　　　長樂鴛鴦　　奏新聲於度曲
　　　妝鳴蟬之薄鬢
30　　照墮馬之垂鬟
　　　反插金鈿
　　　橫抽寶樹
　　　南都石黛　　最發雙蛾
　　　比地燕脂　　偏開兩靨

　　赤有
35　　嶺上仙童　　分丸魏帝
　　　腰中寶鳳　　授歷軒轅
　　　金星將婺女爭華
　　　麝月與嫦娥競爽
　　　驚鸞冶袖　　時飄韓掾之香
40　　飛燕長裾　　宜結陳王之佩
　　　雖非圖畫　　入甘泉而不分
　　　言異神仙　　戲陽臺而無別
　　眞可謂傾國傾城無對無雙者也

　　加以
　　　天情開朗

45 逸思雕華
　　妙解文章
　　尤工詩賦
　　瑠璃硯匣　　終日隨身
　　翡翠筆牀　　無時離手
50 清文滿篋　　非惟芍藥之花
　　新製連篇　　寧止蒲萄之樹
　　九日登高　　時有緣情之作
　　萬年公主　　非無累德之辭
　其佳麗也如彼
55 其才情也如此

　既而
　　椒房宛轉
　　柘館陰岑
　　絳鶴晨嚴
　　銅蠡晝靜
60 三星未夕　　不事懷衾
　　五日猶賒　　誰能理曲
　　優游少託
　　寂寞多閒
　　厭長樂之疏鍾
65 勞中宮之緩箭
　　輕身無力　　怯南陽之擣衣
　　生長深宮　　笑扶風之織錦

　雖復
　　投壺玉女　　爲歡盡於百驍
　　爭博齊姬　　心賞窮於六箸
70 無怡神於眼景
　　惟屬意於新詩

　可得
　　代彼萱蘇　　微蠲愁疾

137

但
　往世名篇
　當今巧製
75　分諸麟閣
　散在鴻都
　不藉篇章
　無由披覽

於是
　燃脂暝寫
80　弄墨晨書
　　　　　撰錄艷歌　　凡爲十卷
　　　　　曾無參於雅須
　　　　　亦靡濫於風人
　涇渭之間　若斯而已

於是
85　麗以金箱
　裝之寶軸
　三臺妙迹　龍伸蠖屈之書
　五色花箋　河北膠東之紙
　高樓紅粉　仍定魯魚之文

90　辟惡生香　聊防羽陵之蠹
　靈飛六甲　高擅玉函
　鴻烈仙方　長推丹沈

至若
　青牛帳裏　餘曲未終
　朱鳥窗前　新妝已竟

方當
95　開茲縹峽
　散此縚繩
　永對翫於書帷
　長循環於纖手

　　豈 如

　　　　鄧 學 春 秋 　　　儒 者 之 功 難 習
100 　　竇 專 黃 老 　　　金 丹 之 術 不 成

　　固 勝

　　　　西 蜀 豪 家 　　　託 情 窮 於 魯 殿
　　　　東 儲 甲 觀 　　　流 詠 止 於 洞 簫
　　變 彼 諸 姬 　　　聊 同 棄 日
　　　　猗 歟 彤 管
105 　　無 或 譏 焉

THE *WEN HSÜAN* AND GENRE THEORY

by
JAMES R. HIGHTOWER

THE *WEN HSÜAN* AND GENRE THEORY

JAMES R. HIGHTOWER

HARVARD UNIVERSITY

The development of genre theory in China has been closely associated with anthology making; the sixth century *Wen hsüan*, with its preface, marks a significant stage in the process. Preliminary to a study of genres in the *Wen hsüan* I shall trace briefly the earlier Chinese interest in the subject, and I might begin by explaining why the subject is significant.

Literary criticism, as distinct from literary theory, is concerned with the individual literary work, which it attempts to understand, interpret, and evaluate. Interpretation and evaluation both depend in part on a true estimate of the writer's intention: he should not be damned for not succeeding in something he never set out to do. What effects he can achieve are limited in the first place by the form in which he puts his composition: within the bounds of a lyric one simply cannot write an epic, nor is a sonnet sequence capable of the kind of effect achieved in a novel. This is, of course, very elementary, and is taken for granted by every practicing critic, whether or not he attempts to formulate his criteria for the several genres. The concept of genre underlies all criticism, and consciously or unconsciously the good critic knows what a given genre is the appropriate vehicle for, as does every competent writer.

But genre is an abstraction; both critic and writer are immediately concerned with the specific, concrete literary work— immediately, but not exclusively. Any literary work is at once an individual entity and a part of a larger whole, that is, the whole corpus of writing which makes up a literary tradition. Genres come into being as a literature develops, not as a single literary work is written. Genre as a class consisting of a single member is meaningless. Consequently even a list of genres has to wait until a literature exists and people are aware of its existence. When they come to abstract the forms in which literature is written, the lists compiled will be a direct reflection of their concept of literature. And conversely, the accepted genres will influence the general idea of what the broader abstraction, literature, is.

We can see this happening in Chinese literary theory from the *Han shu* " Essay on Bibliography " through Ts'ao P'ei and Lu Chi down to Liang times, when genre studies became the first concern of the theorists. The conflict between the traditional didactic view of literature as a vehicle for moral instruction and the heretical view of literature as its own end is exemplified in the divergent lists of literary forms. It was the appearance of the first purely literary genre, the *fu*, which precipitated the controversy, and whether or not it was thought a good thing, the *fu* was recognized as something subject to different standards of evaluation. WANG Ch'ung 王充 (27-?100), for example, in recommending clarity as the ideal of style, could make an exception of the *fu*.[1]

By the end of the second century A. D. the *fu* had become a respectable part of literature. It takes its appropriate place (being the latest arrival on the scene) at the end of the first list of genres, the one in Ts'ao P'ei's " Essay on Literature ": [2]

> Though all writing is essentially the same, the specific forms differ. Thus memorials (*tsou*) and deliberations (*i*) should be decorous; letters (*shu*) and essays (*lun*) should be logical; inscriptions (*ming*) and dirges (*lei*) should stick to the facts; poetry (*shih*) and *fu* should be ornate.

This list served as the model for later ones: the forms are presented in pairs and each pair is characterized by a quality that presumably distinguishes it from all the other pairs.

Ts'ao P'ei was probably not trying to make a complete list of genres; certainly he was not concerned to formulate an exact definition of the ones he did list. His interest in genres was only a by-product of the typical third-century pastime of evaluating and categorizing people. The primary interest was in determining the

---

[1] *Lun heng* (*SPTK* ed.) 30.6a. WANG Ch'ung was not being complimentary: " If abstruse and elegant style makes the sense hard to understand, you have nothing better than *fu* or *sung*." 深覆典雅指意難覩.唯賦頌耳.

[2] 論文, the part of his *Classical Essays* 典論 preserved in *Wen hsüan* (*SPTK* ed., hereafter abbreviated *WH*). For this passage cf. *WH* 52.9a: 夫文本同而末 異.蓋奏議宜雅.書論宜理.銘誄尚實.詩賦欲麗.

fitness of a person for office in terms of his ability and knowledge as against the specific demands made by the office on its incumbent.[3] Extended to writing this process produced something recognizable as literary criticism, but of a rather limited and specialized sort. The first need was to determine what demands writing made on the individual; after that it should be easy to predict the possible performance of a given individual as a writer. A less challenging exercise would be to decide exactly how well any writer had succeeded at his craft.

Ts'ao P'ei may well have been the first to try to work this out on paper. He begins his essay with some general remarks: not all writers are equally good at all kinds of writing; a writer is not an unbiased critic, for he applies as a standard of criticism to others the qualities that enable him to excel in one form, this standard may not be applicable to other forms.[4] In this way Ts'ao P'ei arrived at his list of genres and their dominant characteristics. On the basis of these he felt himself able to pass judgment on the performance of his contemporaries:

> WANG Ts'an excels in *fu*. Hsü Kan, though he sometimes writes in the (vulgar) Ch'i style, is on the whole his equal. The memorials and addresses to the throne of CH'EN Lin and JUAN Yü are the best in this age. YING Ch'ang's are well-balanced but lack strength; LIU Chen's are powerful but not tightly organized. K'UNG Jung has a style and vigor elevated and refined beyond all others, but he cannot sustain an argument—his logic falls short of his stylistic qualities.[5]

The century separating LU Chi's " *Fu* on Literature " from Ts'ao P'ei's " Essay on Literature " is hardly an adequate measure of the enormously increased sophistication of LU Chi's attitude toward the art of letters. But in genre theory there has been little advance; LU Chi has simply added two items (Epitaph and Admonition) to the eight in Ts'ao P'ei's list and substituted Discourse for Letters. His characterizations of each form follow the

---

[3] *Ch'ing t'an* began as an exercise of this sort, and the *Shih-shuo hsin-yü* in part reflects these interests; cf. CH'EN Yin-k'o, *T'ao Yuan-ming's Thought and its Relation to " Pure Talk "* 陳寅恪, 陶淵明之思想與清談之關係 (Harvard-Yenching Institute, 1945), p. 2 ff. See also WANG Yao, *Medieval Literary Thought* 王瑤, 中古文學思想 (T'ang T'i, 1951), p. 132.

[4] *WH* 52.7b-8a.

[5] *Op. cit.*, 52.8b-9a.

same pattern. The epithets are different, but not, on the whole, more appropriate; they are, however, more metaphorical and hence harder to understand:

Lyric poetry (*shih*) traces emotions daintily; Rhymeprose (*fu*) embodies objects brightly. Epitaph (*pei*) balances substance with style; Dirge (*lei*) is tense and mournful. Inscription (*ming*) is comprehensive and concise, gentle and generous; Admonition (*chen*), which praises and blames, is clear-cut and vigorous. Eulogy (*sung*) is free and easy, rich and lush; Disquisition (*lun*) is rarified and subtle, bright and smooth. Memorial to the Throne (*tsou*) is quiet and penetrating, genteel and decorous; Discourse (*shuo*) is dazzling bright and extravagantly bizarre.[6]

What are needed to make this passage intelligible are examples of the things mentioned, to demonstrate how Eulogy, for example, is " free and easy, rich and lush," or at least to show an approved specimen of the form. This lack was supplied by the anthologists, beginning with Lu Chi's contemporary, CHIH Yü 摯虞, who died around 310 A.D.

CHIH Yü compiled the first known anthology of diverse genres and combined with it a statement of the nature and historical development of the genres represented.[7] His *Collection of Writings by Genres* (*Wen-chang liu-pieh chi* 文章流別集) originally contained sixty chapters of texts and was accompanied by two chapters each of Notes 志 and Discussions 論.[8] All that remains of this pioneering effort are a few of his remarks about some of the genres in his anthology. They are very much in the didactic tradition and, except in the treatment of genre, suggest a more limited view of literature than Lu Chi had demonstrated in his *fu*.

CHIH Yü had a strong historical sense, and his obsession with

[6] Achilles FANG, "Rhymeprose on Literature," *HJAS* 14(1951).536. I have transposed the terms and their translations.

[7] There had been earlier anthologies of course. The *Classic of Songs* is after all an anthology of poetry, and its four-fold division into *Feng*, Big and Little *Ya*, and *Sung* may have been an attempt by its compiler to establish different categories of song. The next anthology was the *Ch'u tz'u*, a collection of the specialized verse form known as *sao*, in its present form dating from the second century A.D.

[8] This was the Liang dynasty listing. The *Sui shu* "Essay on Bibliography" (T'ung-wen ed., 35.21a) records a copy in 41 *chüan* and, as a separate entry, *Wen-chang liu-pieh chih lun* in two *chüan*. The two T'ang histories have only *Wen-chang liu-pieh* in 30 *chüan*: *T'ang shu* 47.42a, *Hsin T'ang shu* 60.20b (all Standard Histories cited in T'ung-wen ed.).

the derivation of literary forms from one of the Confucian Classics foreshadows Liu Hsieh's elaborate scheme. For instance, after a passing mention of the Poetry Collectors and the old punning definition of poetry, he finds the various forms of poetry adumbrated in the *Classic of Songs*.[9]

The ancient poetry was in lines of 3, 4, 5, 6, 7, and 9 words. In general the form was a four-word line, but occasionally there would be a line of different length interspersed among the four-word lines. In later times these were extended to whole stanzas (and so developed into separate forms). An example of the three-word line in ancient poetry is " In a flock go the egrets,/ The egrets go flying ";[10] this meter was frequently used in the Han dynasty Suburban Temple Hymns. The five-word line occurs in " Who says the sparrow has no beak? / How else could it have pierced my roof? "[11] It is frequently used in the popular songs of entertainers. An example of the six-word line is " Meanwhile I pour out a cup from that bronze lei-vase ";[12] it is also used in *yüeh-fu* poems. An example of the seven-word line is " ' Kio ' sings the oriole as it lights on the mulberry-tree ";[13] it is used nowadays in the popular songs of entertainers. An example of the nine-word line is " Far away we draw water from that running pool; we ladle it there and pour it out here ";[14] as it does not occur in songs, it is now seldom used.

Derivation of this sort is of course fallacious (though it has been popular in China down to modern times), but it provides a link by which a series of similar literary forms may be brought together under a single, more general head. Carried a bit further it could lead to a single inclusive term for all the different forms of metrical composition, in short, to a word for " poetry." The term *shih* was never stretched quite that far, though Chih Yü was pushing it in that direction, as in the passage where he says,[15] " In later times much poetry (*shih*) has been written; that which praises accomplishments is called Eulogy (*sung*), the rest is collectively known as *shih*." In later anthologies the process was reversed, and the tendency was to set up a separate category for each term, even when the terms were actually synonymous.[16]

[9] Yen K'o-chün's *Collected Chin Prose* 全晉文 77.8b (abbreviated *CCW*), quoted from *I-wen lei-chü* 56 (abbreviated *IWLC*).

[10] 振振鷺，鷺于下 *Mao shih* 298/1, Karlgren's translation.

[11] 誰謂雀無角,何以穿我屋 *Mao shih* 17/2, Waley's translation (p. 65).

[12] 我姑酌彼金罍 *Mao shih* 3/2, Karlgren.

[13] 交交黃鳥止于桑 *Mao shih* 131/2, Waley, p. 311.

[14] 泂酌彼行潦挹彼注茲 *Mao shih* 251/1, Karlgren.

[15] *CCW* 77.7b.

[16] The *Wen hsüan* provides a good example; see note 78 below.

CHIH Yü was also aware of the possibility of a single term standing for two different kinds of writing, the word *sung*, for example. At the end of the passage quoted above he discusses several examples of Eulogy, noting their similarity to the kind of composition which goes by that name in the *Classic of Songs*. Then he says, " Eulogies like those of MA Jung are purely in the modern *fu* form; to call them *sung* is to go far astray." [17] This is the first example, so far as I know, of a Chinese critic who could see the difference between a form and its label.

It is unfortunate that not even the table of contents of CHIH Yü's anthology has survived, so there is now no way of guessing what he considered worth including in a selection of the best of literature as he conceived it. His example inspired a great flurry of anthology making during the next two and a half centuries. The " Essay on Bibliography " in the *Sui shu* [18] lists 419 titles, of which 107, comprising over two thousand scrolls, had survived until early T'ang times. Besides the general anthologies, there are a number of specialized ones—of poetry, of periods, of places, of writings by women, of individuals and of groups of individuals, and finally anthologies devoted to specific genres.

Of all these anthologies only two remain: the *Yü-t'ai hsin-yung* of Hsü Ling 徐陵 (503-583) and HSIAO T'ung's *Wen hsüan*. The *Yü-t'ai hsin-yung* is restricted to love poetry, for the most part in the five-word meter, and so contributes nothing to genre theory. It is of interest as representing the unorthodox tastes of HSIAO Kang 蕭綱 (503-551), under whose patronage it was compiled, and is valuable for preserving a considerable amount of Six Dynasties poetry that otherwise would have been lost.

The *Wen hsüan*, the Anthology par excellence, is a much more catholic collection, containing both prose and verse arranged under thirty-seven separate heads. Its contents are diversified enough to suggest that HSIAO T'ung was trying to provide specimens of all the forms of literature and perhaps even to arrange them in some kind of significant order. His Preface to his an-

---

[17] *CCW* 77.8a.

[18] *Sui shu* 35.21a-26b. CHIH Yü's anthology heads the list and the compiler's comment (*op. cit.*, 35.36a) credits him with providing the model for later anthologists.

thology also includes a list of genres. Unfortunately it is not identical with the list in the table of contents of the *Anthology* itself, nor is the sequence of entries the same. There are anomalies in both lists which are in part resolved by confronting them. I shall begin by translating the Preface,[19] treating in the notes the problems connected with the list of genres.[20]

When we look to the first beginnings and scrutinize from afar those primordial conditions—in times of winter caves and summer nests when men devoured undressed game and drank blood [21]—then times were rude and people plain: writing [22] had not yet appeared. Then we come to the rule of Fu-hsi, who first traced the Eight Trigrams and invented writing to take the place of government by knotted cords; from this time written records came into being.[23]

The *I ching* says,[24] " Observe the patterns in the sky to discover the seasons' changes; observe the patterns among men to transform All-Under-Heaven " —so far-reaching are the times and meanings of pattern (*wen*)! [25] Now the Imperial Chariot had its origin in the oxcart,[26] but the Imperial Chariot has

---

[19] There is a translation by G. MARGOULIÈS, Le " fou " dans le Wen-siuan (Paris, 1926), pp. 22-30. I have made use of the commentaries brought together in KAO Pu-ying's Wen-hsüan Li chu i-su 高步瀛，文選李注義疏, Peking, 1934.

[20] It should be pointed out that this Preface is not simply a straightforward piece of expository prose. It is written in Parallel Style, and logical exposition frequently gives way before the demands of symmetry.

[21] " Formerly the ancient kings had no houses. In winter they lived in caves which they had excavated, and in summer in nests which they had framed. They knew not yet the transforming power of fire, but ate the fruits of plants and trees, and the flesh of birds and beasts, drinking their blood, and swallowing (also) the hair and feathers." (*Li chi* 9/5, James LEGGE, *Sacred Books of the East*, 27.369.)

[22] 斯文. This first occurrence of the protean word *wen* gets its overtones of " culture," " civilization " from its association with *Lun yü* 9/5: 文王旣沒,文不 在茲乎，天之將喪斯文也，後死者不得與斯文也 , but here the emphasis is on " written texts."

[23] This is quoted verbatim from the opening lines of the " Preface " to the *Shu ching* attributed to K'ung An-kuo. The text is in *WH* 45.31b. The more commonly accepted tradition (given in Hsü Shen's Preface to the *Shuo wen*) says it was Shen-nung, Fu-hsi's successor, who " ruled by knotted cords," and credits Ts'ang-chieh, Huang-ti's minister, with the invention of writing, based on his observation of " the tracks made by the feet of birds and animals."

[24] *I ching* No. 22, LEGGE (*SBE* 16), 231.

[25] Cf. *op. cit.*, No. 16, where the same encomium is pronounced of the hexagram *yü*: 豫之時義大矣哉 " Great indeed are the time and significance indicated in Yü! " (*ibid.*, 227). This formula occurs also of hexagrams 17, 33, 44, and 56.

[26] The commentators identify 椎輪 with 椎車 of *Huai-nan tzu* 17.4b (*SPTK* ed.): 古之所爲不可更則椎車.至今無蟬⊙ (罤 + Rad. 22). " Something which has not changed since ancient times is the oxcart; it still has no spokes," i. e., it has solid

none of the crudeness of the oxcart. Thick ice is composed of accumulated water, but accumulated water has not the coldness of thick ice. Why so? The original form is preserved but elaborated on, or the essential nature changed through intensification. This is true of things, and it is also true of literature (*wen*). It changes with passing time, and to describe it is no easy task. But to make the attempt:

The Preface to the *Classic of Songs* says,[27] " There are six modes of the Songs. The first is instruction (*feng*), the second is description (*fu*), the third is simile (*pi*), the fourth is metaphor (*hsing*), the fifth is ode (*ya*), the sixth is hymn (*sung*)." Later poets deviated from the ancient [practice], and of the [six modes of the] ancient poetry, the moderns took over only the term *fu*. It appeared first of all in the works of Hsün Tzu [28] and Sung Yü,[29] and was continued subsequently by Chia I [30] and Ssu-ma Hsiang-ju; [31] from this

---

wheels. The development of the imperial chariot from the oxcart as an example of simple origins for elaborate devices appears already in Lu Chi's *fu* " The Feather Fan " 羽扇賦 (*CCW* 97.4b): 玉輅基於椎輪. (Tseng Chao 曾釗).

[27] *Mao shih* Preface, text in *WH* 45.30a-b. This incongruous list first appeared in the *Chou li* 6.13a (*SPTK* ed.). It is hopeless to translate the terms satisfactorily, for they have meant many things to different commentators, but at least the nature of the difficulty can be defined. Three of the six items (*feng, ya, sung*) are the names of the chief divisions of the present *Classic of Songs*, and while there is no general agreement about their significance there, they are certainly not the names of tropes. *Fu, pi,* and *hsing* are variously interpreted and inconsistently applied by the commentators on the *Classic of Songs*. The *Mao shih* Preface says nothing about them, and it is likely that its author was quoting the traditional list without bothering about the extraneous items. The sequence may be a clue to how the list was originally understood, and Tseng Chao suggests that *feng* is to be taken with *fu, pi,* and *hsing*, and hence not as the name of the first section of the *Shih ching*. For present purposes the important question is how Hsiao T'ung understood the items, and it is apparent from the rest of this paragraph that he was concerned solely with the occurrence of the word *fu* as something associated with the *Classic of Songs*. It provides his point of departure in sketching the development of the *fu* genre, though he must have been aware that the genre was not identical with the trope, as indeed his statement in the next sentence " the moderns took over only the term *fu* " implies. The association was not original with him; Pan Ku (Preface to the " Two Capitals *fu*," *WH* 1.1a) and Chih Yü (*CCW* 77.8a) both asserted the derivation of *fu* from the *Classic* in the same words: 賦者古詩之流也.

[28] The riddles in rhyme of the " Fu p'ien " of *Hsün tzu* have nothing in common with the *fu* genre of Han times.

[29] Four *fu* attributed to Sung Yü (3rd century B.C.) are included in *WH* (13.1a, 19.1a, 19.8b, 19.12a); six more are in *Ku wen yüan 2.*

[30] Chia I's " Owl *fu* " (*WH* 13.20b) is the earliest *fu* of which the text is given in a contemporary Former Han source (*Shih chi* 84.10b-13b).

[31] Of the three *fu* attributed to Ssu-ma Hsiang-ju in *WH* (7.23b, 8.1a, 16.10a), two (" Tzu-hsü " and " Shang-lin ") are quoted in his *Shih chi* biography (*Shih chi* 117.12b-26b). The authenticity of the third (" Ch'ang-men " 長門) has been questioned (e. g., by Ku Yen-wu, *Jih-chih-lu* 19.11a, Sao-yeh Shan-fang ed.) because of

time on the ramifications were many. Descriptive of cities and sites there are [the *fu* of CHANG Heng and SSU-MA Hsiang-ju with their imaginary inter-locutors His Honor] Insubstantial [32] and [Master] No-Such[-Person].[33] Directed against hunting are the "Ch'ang-yang"[34] and "Hunting with Plumes"[35] [*fu* of YANG Hsiung]. When it comes to *fu* describing one event or celebrating a single object (such as those on Wind, Clouds, Plants, and Trees, or the ones about Fish, Insects, Birds, and Beasts), considering their range, it is quite impossible to list them all.[36]

There was also the Ch'u poet CH'Ü Yüan, who clung to loyalty and walked unsullied; the prince would not accept it,[37] when the subject offered advice unwelcome to his ears.[38] Though his understanding was profound and his plans far-reaching, in the end he was banished south of the Hsiang River. Injured for his unbending integrity and with no one in whom to confide his sorrow, he stood on the verge of the abyss, determined to embrace the stone; he sighed by the pool, haggard in appearance.[39] It is from him that the writings of the *sao* poets derive.

Poetry is the product of the emotions: the feelings are moved within and take form in words.[40] In "The Osprey" and "The Unicorn" appears the Way of the Correct Beginning; [41] "The Mulberry Grove" and "On the Banks

---

the anachronism in the preface and the stylistic differences between it and the surely genuine *fu* of SSU-MA Hsiang-ju. Neither is conclusive ground for rejecting the "Ch'ang-men *fu*," but the fact that it is not mentioned in either the *Shih chi* or the *Han shu* biography, despite the romantic setting, makes it highly suspect.

[32] 憑虛公子 is a character in the "Hsi ching *fu*" of CHANG Heng (*WH* 2.1a).

[33] 亡是公 occurs in SSU-MA Hsiang-ju's "Shang-lin *fu*;" as CHANG Shao 張杓 remarks, HSIAO T'ung places it in the subcategory of Royal Hunts (*WH* 8.1a, where it belongs) rather than with the Capitals *fu* where CHANG Heng's "Hsi ching" is found.

[34] 長楊, after the palace of that name at Ch'ang-an, where the game was brought in cages and released. (*WH* 9.1a).

[35] 羽獵, *WH* 8.20b.

[36] A glance at the table of contents of the *Li-tai fu-hui* 歷代賦彙 (which is arranged by categories) will show that this is no exaggeration.

[37] 從流, lit., "follow the current," to be understood as an allusion to *Tso chuan* (Chao 13), "Duke Huan of Ch'i followed the good as a current" 齊桓從善如流 (KAO Pu-ying).

[38] 逆耳. The proverbial expression 良藥苦口,利於疾.忠言逆耳,利於行 "Good medicine is bitter to the mouth but is of benefit to one who is sick; loyal words offend the ears but are of benefit to the conduct" appears (with minor variations) in *K'ung-tzu chia-yü* 4.1b and *Shih chi* 55.4a (KAO Pu-ying).

[39] Paraphrased from CH'Ü Yuan's biography in *Shih chi* 84.4b.

[40] *Mao shih Preface* (*WH* 45.30a): "Poetry [詩 i. e., the *Classic*] is the product of the emotions. In the heart it is emotion; expressed in words it is poetry. The feelings are moved within and take form in words."

[41] 關雎,麟趾, *Shih* Nos. 1 and 11.

of the Pu " represent the music of a defunct state.⁴² Truly the way of the
Feng and the *Ya* may be seen in them at its most brilliant.⁴³ From the middle
period of Fiery Han ⁴⁴ the paths of poetry gradually diverged. The Retired
Tutor (Wei Meng) wrote his " Poem in Tsou," ⁴⁵ and the surrendered general
(Li Ling) wrote the poem on the bridge; ⁴⁶ with them the four-word and
five-word [meters] became [recognized as] distinct classes. In addition there
were [meters] with as few as three words and as many as nine words, the
several forms developing at the same time, [like horses] galloping together
though on separate traces.

Eulogy (*sung*) serves to broadcast virtuous deeds; it praises accomplish-
ments.⁴⁷ Chi-fu made his pronouncement " How stately! "; ⁴⁸ Chi-tzu exclaimed " Oh perfect! " ⁴⁹ Elaborated as poetry it was expressed like that;
composed as eulogy it is also this way.⁵⁰

---

⁴² 桑間,濮上 : cf. *Li chi* 19/1: " The notes of the ' Sang chien ' and ' Pu shang '
are those of a defunct state." (Couvreur 2.49).

⁴³ The first two are easily understood as representing the " way of the *Feng* at its
most brilliant " (they are singled out by the *Mao shih* Preface: 關雎麟趾之化，
王者之風), but it is not clear why the " Sang-chien " and " Pu-shang " should stand
for the *Ya*, since these are not even among the six lost *Songs*.

⁴⁴ The Han ruled by virtue of the Fire element.

⁴⁵ Wei Meng 韋孟 (second century B.C.) was Tutor to three generations of
Princes of Ch'u, the last of whom he found intractab!e and against whom he " wrote
a satirical poem as a remonstrance " 作詩風諫. He retired to Tsou (his native
place), where he wrote another poem, presumably the one referred to by Hsiao T'ung,
though it is only the first that he included in his anthology (*WH* 19.24a). Both poems
are quoted in *Han shu* 73.1a, 3b, followed by Pan Ku's remark, " Some say these
poems were written by one of [Wei Meng's] descendants who tried to give expression
to his ancestor's feelings." (*Op. cit.*, 73.4b).

⁴⁶ Referring to the farewell poem to Su Wu (*WH* 29.12b), which begins " We clasp
hands on the river bridge / By nightfall where will the traveler have gone? " It is
now generally accepted that all the Li Ling poems in the *Wen hsüan* are forgeries.

⁴⁷ Cf. *Mao shih* Preface: " Eulogy praises the manifestations of flourishing virtue,
and announces its accomplishment to the spirits." (*WH* 45.31a).

⁴⁸ *Mao shih* No. 260/8: 吉甫作誦,穆如清風 " Chi-fu has made this eulogy,/
Stately its clear melody." This poem is a eulogy of Chung Shan-fu, but it is not in
the " Sung " section of the *Shih*; neither is the preceding eulogy of the Prince of Shen
(*Mao shih* No. 259/8) with its similar concluding lines: 吉甫作誦,其詩孔碩.其
風肆好,以贈申伯. However, in both cases what Chi-fu is praising is the song
itself, and Hsiao T'ung probably read 誦 as 頌, which may have been an alternative
reading in other texts of the *Shih* than Mao's (cf. Ch'en Huan 陳奐,詩毛氏傳疏
6.65, Basic Sinological Series ed.).

⁴⁹ Chi-tzu is the " Duke's-son Chao of Wu " who came on a visit of state to Lu
(*Ch'un ch'iu*, Hsiang 29). The *Tso chuan* gives a long account of his reception,
particularly of the musical performance which he requested and which included
selections from the major sections of the *Classic of Songs*. After each piece he made
appropriate remarks. His exclamation " How perfect " came after he had heard the
Sung section of the *Shih* and is followed by an enthusiastic catalog of its perfections;
cf. Legge, *The Chinese Classics* 5.550.

Next are Admonition (*chen*),[51] which arises from making good defects, and Warning (*chieh*),[52] which derives from setting to rights. Disquisition (*lun*) is subtle in making logical distinctions,[53] and Inscription (*ming*) is generous in narrating events.[54] When a good man dies, a Dirge (*lei*)[55] is made; when a portrait is painted, an Appreciation (*tsan*)[56] is supplied.

---

[50] 舒布爲詩,既言如彼.總成爲頌,又亦如此 . This ambiguous and unilluminating line has been variously interpreted by the commentators. The antecedents of the "that" and "this" are vague at best. I should like to take them as referring to the remarks of Chi-tzu and Chi-fu, respectively. Since Chi-tzu was not composing a formal eulogy, but was eulogizing the Eulogies (*Sung*) of the *Classic of Songs*, and Chi-fu was pronouncing a eulogy of his own eulogy (which strictly speaking is a Song [*shih*], as it does not appear among the Eulogies [*Sung*] proper), what HSIAO T'ung characterizes as *shu-pu wei shih* and *tsung-ch'eng wei sung* could be the objects of their remarks as he quoted them, i. e., a Song (*shih*) of the *Shih ching* and the *Sung* section of the *Shih ching*.

However, KAO Pu-ying says "that" refers to the *sung* section of the *Shih*, and "this" to the modern eulogy, which is a special form derived from the *sung* of the *Shih* in the same way modern *fu* is derived from the *fu* of the Six Modes of the *Shih*. The objection to this interpretation is the lack of any previous mention of "modern eulogy" as separate from the *Shih* eulogies. Whatever HSIAO T'ung had in mind, he was obviously more concerned with rhetorical symmetry than unambiguous statement.

[51] 箴 . The sole specimen of this rhymed genre in *WH* (56.1a) is CHANG Hua's 張華 (232-300) "Admonition to the Lady Recorder" 女史箴, a homily on the Confucian ideal of female behavior ("All know enough to adorn their faces, but none enough to adorn her conduct") addressed to the legendary lady officer who was supposed to record the acts of the Empress and (according to the commentators) directed against the influence of the Empress' clan. The classic examples of Admonition are those by YANG Hsiung entitled "Admonitions to the Twelve Provinces and the Twenty-five Officers" 十二州二十五官箴, of which (according to YEN K'o-chün, *Ch'üan Han wen* 54.9a, abbreviated *CHW*) only twenty-eight survive.

[52] 戒 . There are no examples in *WH*. LIU Hsieh (*Wen-hsin tiao-lung* 4.51a-b, FAN Wen-lan ed., abbreviated *WHTL*) appends a paragraph on Warning to his chapter on 詔策, where he refers to Han Kao-tsu's testamentary charge to his Heir Apparent 手敕太子文 (*Ku wen yüan* 10.1b-2b, *SPTK* ed.), TUNG-FANG So's warning to his son 誡子詩 (*Tung-fang Ta-chung chi* 37a, *Po-san-chia chi* ed.) and PAN Chao's "Precepts for Women" 女誡 (translated by N. L. Swann, *Pan Chao, Foremost Woman Scholar of China*, pp. 82-90).

[53] 論則析理精微: this is modified from LU Chi's "Wen fu": 論精微而朗暢 "*Lun* (disquisition) is rarified and subtle, bright and smooth" (Achilles FANG, *op. cit.*, 536). *WH* 51-55 has thirteen specimens of *lun*, beginning with CHIA I's "Critique of Ch'in" 過秦論. In Chs. 49-50 are nine examples under the generic heading "Disquisitions from the Histories" 史論. Most are from FAN Yeh's *Hou-Han shu*; the one from PAN Ku's *Han shu* is the Appreciation 贊 appended to the "Biography of Kung-sun Hung," which might be expected to appear under the next *WH* heading, "Appreciations from Narratives in the Histories" 史述贊.

[54] 銘則序事清潤. I have emended 清 to 溫 to agree with the "Wen fu": 銘博約而溫潤 "*Ming* (inscription) is comprehensive and concise, gentle and

generous" (FANG, *op. cit.*, 536). *WH* 56 has five *ming*, a very miscellaneous group of compositions. The term applies to any text that is inscribed or carved on anything, from the insistent motto on Shun's bathtub (苟日新,日日新,又日新 *Li chi* 42/1) to the patent of enfeoffment cast on a bronze tripod or the eulogy on a tombstone. Only confusion results from its use as the name of a literary genre. Both LU Chi and LIU Hsieh associate *ming* with *chen*, because the *ming* engraved on a weapon or utensil usually is a rhymed Admonition about its proper use. For instance, TS'AI Yung (132-192) is credited with a Beaker Inscription 樽銘 (*IWLC* 73.10b) warning against excess in drinking, while YANG Hsiung wrote a Warning about Wine 酒箴 (*Han shu* 94.11a-b) which is actually in praise of drinking. LI Yu 李尤 (*ca.* 55-135) wrote a series of eighty-five *ming* to be inscribed at passes and fords, on gates and pillars, on screens, swords, musical instruments, ink stones, brushes, slippers, weapons, tables, etc. (*Ch'üan Hou-Han wen* 50.4a-13a, abbreviated *CHHW*). All of them are short and rhymed. But a Grave Inscription (*pei*) often concludes with a *ming*, which is a rhymed eulogy of the deceased, and so *ming* becomes confused with *sung*.

The first of the *ming* in *WH* is by PAN Ku (32-92), written on the occasion of a sacrifice to Mt. Yen-jan. It is very short (five lines) in a *sao* meter, but is preceded by a long prose introduction (celebrating TOU Hsien's 竇憲 campaign), part of which is also rhymed.

The second in TS'UI Yuan's 崔瑗 (77-142) "Inscription on a Warning Vessel" 座右銘 advising moderation and written in rhymed prose. The third is CHANG Tsai's 張載 (*ca.* 290) "Inscription on Sword-Gate Pass," which might be a *fu*, except that its descriptive passages are more restrained. It supplied LI Po with one line for his "The Way to Shu is Difficult" 蜀道難 (*Works* 3.4b, *SPTK* ed.).

The author of the two remaining *ming* in *WH* is LU Ch'ui 陸倕 (470-526). One is the "Inscription on Stone Gate," a rhymed four-word eulogy of the Liang Emperor Wu, who ordered the piece; it is introduced by a prose account of the Emperor's accomplishments. The "Inscription for a New Waterclock" 新漏刻銘 was also written to order; in it too the *ming* proper follows a long unrhymed preface in Parallel Prose.

[55] There are eight 誄 in *WH* 56-57. All are in a rhymed four-word meter with prose preface. (Some irregular lines occur in the single *lei* by TS'AI Yung [*Works* 9.6a, *SPTK* ed.].) LIU Hsieh (*WHTL* 3.13b) discusses *lei* together with *pei*. There is no formal difference between some *lei* and *pei*; presumably it was the use (i. e., the latter was engraved on stone) that determined the name. In fact one *lei* (YANG Hsiung, "Dirge for the Empress Yüan," *CHW* 54.9b) uses the term *ming* for the rhymed part of his *lei*. CHIH Yü (*Wen-chang liu-pieh lun, CCW* 77.9b) says, "There are examples from antiquity of *shih, sung, chen,* and *ming* which can be used as models for such compositions. But for the *lei* there is no established form, and so there is much diversity among their authors."

[56] There are two 贊 in *WH* 47. The first is a eulogy (it is introduced as *sung*), by HSIA-HOU Chan 夏侯湛 (243-291), of TUNG-FANG So, inspired by seeing his portrait, but YÜAN Hung's 袁宏 (328-376) "Appreciation of Famous Ministers of the Three Kingdoms" 三國名臣序贊 which follows makes no mention of any portraits. Both have prose prefaces, and the *tsan* proper is in four-word rhymed verse. The distinction between *tsan* and *sung* is hazy at best. SUN Ch'u 孫楚 (d. 293), for example, writes a "Eulogy of Confucius" 尼父頌 and an "Appreci-

Further there are these branches:[57] Proclamation (*chao*),[58] Announcement (*kao*),[59] Instruction (*chiao*),[60] and Command (*ling*);[61] these types:[62]

---

ation of Yen Hui" 顏回贊 (*CCW* 60.7b, 8a). The four-word rhymed Appreciations appended to *Hou-Han shu* biographies which HSIAO T'ung assigns to a special category ("Appreciations from Narratives in the Histories"; cf. note 53) are formally identical with *tsan*, as are the *tsan* which follow each section of the *WHTL*. This sort of summary and judgment derives of course from the estimates in prose with which SSU-MA Ch'ien concluded each chapter of his *Shih chi*, introducing each with the phrase 太史公曰.

[57] No importance should be attached to the collective terms 流,列,品,作,制,文 which are all interchangeable in this context and best left untranslated. They function simply to divide the list into groups, some of which are obviously of related items. However, these groupings also raise serious problems, particularly when the type of composition named is not represented in the anthology itself. The difficulties are discussed in the following notes, but I am not satisfied that any are satisfactorily solved. The question of omissions is also pertinent (why 表 and 奏, but no 啟 and 疏, both of which appear in the anthology?). These omissions may supply the necessary clue, namely, that HSIAO T'ung compiled this list, not with the idea of system and completeness, but to achieve rhetorical symmetry. This is not to accuse him of carelessness, but rather suggests that he may have had other considerations in mind, euphony perhaps. This first group is relatively homogeneous, in that each type of composition is directed from a superior to inferiors.

[58] The two examples of 詔 in *WH* 35 are both by the Han Emperor Wu, the first to provincial officials ordering them to recommend good men to the throne, the second concerning the examination of these same. According to LIU Hsieh (*WHTL* 4.50b), *chao* is addressed by the Emperor to his officers; in Han times its drafting was the responsibility of the Masters of Documents 尚書. In *WH* 35, *chao* are followed by Patents of Enfeoffment 冊 (LIU Hsieh writes 策), a category not mentioned in the Preface.

[59] 誥 does not appear in *WH*. The prototypes are the several Announcements in the *Shu ching*. CHANG Heng (78-139) wrote an "Announcement concerning the Emperor's Tour of the East" 東巡誥 (*CHHW* 54.3a). LIU Hsieh does not discuss *kao* as a separate category.

[60] 教, of which *WH* 36 contains two examples, both by FU Liang 傅亮 (d. 426), are instructions emanating from a noble (LIU Hsieh, *WHTL* 4.51b). Usually some action is called for by an Instruction; both of FU Liang's deal with the restoration of shrines.

[61] The only 令 in *WH* (36) is one written by JEN Fang (460-508) on behalf of the Ch'i Empress Hsuan-te, urging HSIAO Yen to accept a patent as Duke of Liang. The *WH* commentator LIU Liang 劉良 says that *ling* was the term used for orders issued by an empress or heir apparent in Ch'in times. However, the series of *ling* issued by TS'AO Ts'ao and his sons TS'AO P'ei and TS'AO Chih are mostly in the form of exhortation and explanation of policy. Some are orders: "Search out able men" (求賢令, Ts'AO Ts'ao, *Ch'üan San-kuo wen* 2.7a, abbreviated *CSKW*); "Let their taxes be remitted for two years" (TS'AO P'ei, *op. cit.*, 6.7a); and others refute wrong ideas: "Promote the able regardless of status" (舉賢勿拘品行 令, *op. cit.*, 2.12a). One of TS'AO Ts'ao's *ling* is a long autobiographical disquisition

Memorial (*piao*),[63] Proposal (*tsou*),[64] Report (*chien*),[65] and Memorandum (*chi*);[66] these categories: Letter (*shu*),[67] Address (*shih*),[68] Commission

___

on why he will not usurp the throne (自明本志令, *op. cit.*, 2.7b); in another he requests advice (求言令, *op. cit.*, 2.4a). All *ling* cite precedents and examples, and though some include an order, a more descriptive translation of the term would be "Policy Statements."

[62] Besides 表, memorials to the throne are also called 章, 上書, 啟, 疏, and 奏. These "types" are distinguished by criteria which are neither consistent nor mutually exclusive. Some are defined by their subject matter: "*Piao* sets forth a request" (*WHTL* 5.9b, where a variant reading 情 for 請 gives "In *piao* one expresses his feelings"). "*Chang* expresses thanks for favors received" (*ibid.*). "*Shang shu* is a generic term for business communications addressed to a ruler" (*ibid.*). "*Tsou* reports on investigations" (*ibid.*). "*Tsou* is the inclusive term for statements on government affairs, precedents, crises, impeachments" (*op. cit.*, 5.19b). "Since Han times an alternative term for *tsou shih* 奏事 was *shang shu* 上疏" (*ibid.*). "As *ch'i* concerns government affairs, it is a species of *tsou*; as it is used to decline rank or express thanks for favors, it is a kind of *piao*" (*op. cit.*, 5.20b). Another set of criteria is purely formal, depending on the stereotyped phrases with which the communication opens and closes. (Ts'AI Yung 蔡邕, *Tu tuan* 獨斷 A.4b-5a, *SPTK* ed., lists some of these.)

[63] There are nineteen *piao* in *WH* 37-38. They cover a variety of topics: three are letters of recommendation, seven are to decline a post or title, four are requests for posthumous titles or the care of graves, one asks for a job, one asks for leave of absence, etc. The range of subject matter can be extended, if one looks beyond the *WH* selection, to include remonstrances, objections, and proposals of all sorts. Their formal opening always uses the phrase "Your subject states" or "Your subject (so and so) has heard" 臣言, 臣 (某) 聞. Exceptionally, one is preceded by the more elaborate formula associated with *chang* or *tsou*: "Your subject so and so, striking his head on the ground, guilty of a capital crime, offers this communication. Your subject has heard that. . . ." (LIU K'un's "Memorial Requesting [SSU-MA Jui] to Ascend the Throne" 劉琨, 勸進表, *WH* 37.31b). The formal close is less stereotyped, but usually says in effect, "Your subject presents his memorial in fear and trembling and humbly requests Your Majesty to pay attention to it."

[64] There is no category labeled *tsou* in *WH*. Numerous examples in the *Po san chia chi* collections have in common with *piao* the same formal opening but lack any regular close; see note 63.

[65] 牋 is usually a letter addressed to a superior. *WH* 40 has nine examples, none to a ruling emperor (this distinguishes them from *piao* and *tsou*), and none to an equal or inferior of the writer (thus differing from *shu*). Ts'AO Chih's letter to YANG Hsiu (*WH* 42.17b) is *shu*; YANG Hsiu's reply (*WH* 40.16a) is *chien*. Outside the *WH* examples usage is not so consistent. K'UNG Jung addresses several *shu* to Ts'AO (*CHHW* 83.7b-8b), and FU Hsien 傅咸 (239-294) addresses both *shu* and *chien* to the Prince of Ju-nan (*CCW* 52.8b, 9a).

[66] There is no example labeled simply 記 in *WH*. There is JUAN Chi's 奏記 (*WH* 40.38a), declining office and addressed to the minister CHANG Chi, which is indistinguishable from *chien*. LIU Hsieh distinguishes between the recipients of these communications: *chi* is addressed to a capital official, *chien* to a provincial general (or

(*fu*),[69] and Charge (*chi*);[70] these compositions:[71] Condolence (*tiao*),[72] Requiem (*chi*),[73] Threnody (*pei*),[74] and Lament (*ai*);[75] these forms: Replies

---

governor?) (*WHTL* 5.41b), but this is contradicted by practice. Other compositions termed *chi* are better called "records" or "description," as Chu-ko Liang's account of the Huang-ling Temple 黃陵廟記 (*CSKW* 59.8a-b), of questionable authenticity, and of course the famous descriptive pieces of Liu Tsung-yüan are *chi*, but this is a T'ang dynasty usage.

[67] 書. *WH* 41-43 gives the texts of twenty-four letters, all addressed to equals or inferiors. There are a few traces in these letters of what was to become a distinctive epistolary style, though no stereotyped opening and close is common to them all. The letter of Ch'iu Ch'ih 丘遲 (464-508) to Ch'en Po-shih, written after the latter's decease (*WH* 43.22b) is something of an anomaly, and Liu Hsin's 劉歆 (d. 23 A.D.) " Letter to the Doctors of the T'ai-ch'ang-ssu " (*WH* 43.29b) is less a personal letter than a public defense of his texts of the Classics. Although I have translated *shu* as " Letter," since all the *WH* examples are that, its occurrence in this present list suggests the meaning " contract "; cf. *WHTL* 6.24a, where there is a similar list: 符檄書移.

[68] There is no 誓 category in *WH*. All but the last of the six *shih* in the *Classic of Documents* are addressed to troops before a battle, and the last, according to one tradition, was also delivered to the survivors of a defeated army. Liu Hsieh defines *shih* as " instructions to troops " (*WHTL* 4.50a), and associates it with 命 " Command "; elsewhere (*op. cit.*, 2.75a) he mentions *shih* as a species of covenant 盟. From its association here with *fu* and *hsi*, the *Shu ching* usage would appear to be the one intended. However, this leaves *shu* a complete anomaly in this group.

[69] 符 does not occur in *WH* as a separate category. The three examples under the heading 符命 " Investiture with the Mandate " (*WH* 48) are a special type in praise of the legitimacy of a dynasty (hence placed after Appreciations?), which Liu Hsieh (*WHTL* 5.1a) treats as 封禪 " Essays on the *feng* and *shan* Sacrifices." *Fu* as Commission appears in *WHTL* 5.42b, also 6.24a; see note 67.

[70] There are five examples of 檄 in *WH* 44; probably the 移文 of K'ung Chih-kuei 孔稚圭 (447-501) in the preceding chapter (category 書) also belongs here, for *hsi* and *i* are similar types (*WHTL* 4.62b treats them together in sec. 20).

[71] The following four forms are closely related, all commemorating a deceased person. *Tiao* and *ai* are indistinguishable formally, but Liu Hsieh (*WHTL* 3.31a-b) defines them in terms of subject: *ai* is written for one who dies prematurely, *tiao* for one who has failed to realize his ambition. Both are rhymed. *Chi* is further distinguished by its use as a part of a mourning ritual; it accompanies a sacrifice to the dead. *Pei* as a separate form is not mentioned in *WHTL* or elsewhere, so far as I can discover.

[72] There are two examples of 弔 in *WH* 60. Chia I's " Condolence for Ch'ü Yüan " is appropriately in the *sao* form. Lu Chi's " Condolence for Ts'ao Ts'ao " has a long prose preface; the *tiao* proper is in a six-word *fu*-type verse. Both pieces are written about notable persons of the past. The emphasis is on the frustrations they suffered, not the grief of the author (see *chi* 祭 below); they are not simply eulogies (*lei, sung, tsan, pei*), and may be critical of the subject's conduct.

[73] *WH* 60 has three 祭. The " Requiem " by Yen Yen-chih 顏延之 (384-456) for Ch'ü Yüan differs from Chia I's " Condolence " only in using a regular four-word

to Opponents (*ta k'o*) [76] and Evinced Examples (*chih shih*); [77] these texts: Three Word (*san yen*) and Eight Character (*pa tzu*); [78] Song (*p'ien*), Elegy (*tz'u*), Ditty (*yin*), and Preface (*hsü*); [79] Epitaph (*pei*) and Columnar

line in place of the irregular *sao* verse. HSIEH Hui-lien's (394-430) "Requiem at an old Grave" is nearer the usual form in that it closes with the stock lament 嗚呼哀哉, but like the "Requiem for Ch'ü Yüan" it is addressed to the spirits of someone not personally known to the writer. (The occasion was the re-burial of two sets of bones.) The "Requiem for Yen Yen-chih" by WANG Seng-ta 王僧達 (423-458) is the only one of the three which is altogether typical of the form, being written to accompany a sacrifice to the spirit of an acquaintance who has just died (the piece is dated in the year of YEN Yen-chih's death).

[74] I have been unable to identify 悲 as a literary form; hence, the translation is purely fanciful. It is perhaps included to complete the phrase.

[75] *WH* 57-58 has three examples of 哀. The first is P'AN Yüeh's moving "Lament for the Eternally Departed," on the occasion of his wife's death. It is written in a short line *sao* meter, with rhyme. The other two are labeled 哀策文; both are for the burial of deceased Empresses, one being a re-interrment. These latter were also written for other members of the royal family than empresses, and Laments generally are not restricted to women, as one might assume from these examples; e. g., the fragment of PAN Ku's "Lament for General Ma Chung-tu" (*CHW* 26.9a).

[76] There is no category in *WH* labeled either 答客 or 指事. TUNG-FANG So's "Answer to a Visitor's Objections" 答客難 (*WH* 45.2b) may be taken as an example of the former. It appears under the general heading "Essays on Set Subjects" 設論 along with YANG Hsiung's "Justification in the Face of Ridicule" 解嘲 and PAN Ku's "Reply to a Guest's Mockery" 答賓戲. The preceding category 對問 "Dialog," with SUNG Yü's "Dialog with the King of Ch'u" as the only specimen, could easily be used to cover all of these.

[77] There is no convincing explanation of 指事. Lü Yen-chi says YANG Hsiung's "Justification" (*WH* 45.7a) is an example. It appears among the "Set Essays" (see note 76), and KAO Pu-ying thinks it is too much like the "Replies to Opponents" to justify a separate designation. He agrees with TSENG Chao that HSIAO T'ung was here alluding to his category of "Sevenses," (of which MEI Sheng's "Seven Persuasions" 七發 [*WH* 34.1a] is the prototype), since in that form a series of seven situations are cited to divert the indisposed or indifferent auditor. In meter and form the "Sevenses" are indistinguishable from Han dynasty *fu* of the *Tzu-shü / Shang-lin* type, and in *WH* they follow *sao*. The term *chih shih* is used by LIU Hsieh in characterizing various forms of writing, and always in the sense of "to cite concrete examples," as for example "Memorials 表章 at the beginning of the Wei were concrete 指事 and factual, and if you are looking for elegance, are not praiseworthy." (*WHTL* 5.10a).

[78] Speculation about 三言 and 八字 is plentiful and inconclusive. There exist three-word and eight-word meters among *yüeh-fu* poetry (the latter is a rarity), but this hardly seems the place for Hsiao T'ung to bring them up.

[79] 篇辭引序. This is another enigmatic group. *P'ien* is a generic name for the sections of a book; it also occurs in *yüeh-fu* song titles. *Yin* likewise is used to mean "song" or "tune." Its Sung dynasty meaning of "Preface" is tempting but altogether anachronistic. *Tz'u* was common in Han dynasty writings as a synonym for

Inscription (*chieh*);[80] Necrology (*chih*)[81] and Obituary (*chuang*).[82] A multitude of forms have shot up like spear-points; diverse tributaries have joined the main stream. Yet they might be compared to musical instruments made of different materials—some of clay, some from gourds, yet all are to give pleasure to the ear; or to embroideries of different colors and designs: all are to delight the eye. This accounts for just about all that writers have written.

When not busy with my duties as Heir Apparent,[83] I have spent many idle days looking through the garden of letters or widely surveying the forest of literature, and always I have found my mind so diverted, my eye so stimulated, that hours have passed without fatigue. Since the Chou and the Han, far off in the distant past, dynasties have changed seven times and some thousands of years have elapsed. The names of famous writers and men of genius overflow the green bag,[84] the scrolls of winged words and flowing brushes fill the yellow covers. If one does not leave aside the weeds and

---

fu or *sao* (cf. the title *Ch'u tz'u*), or even "poetry" generally. *WH* 45 actually has a category *tz'u*, containing two disparate compositions: the "Autumn Wind" 秋風 of the Han Emperor Wu and T'ao Ch'ien's "Return" 歸去來辭. The first is a "Ch'u Song" and logically should be under "*yüeh-fu*" or the subhead 雜歌 of *shih*, along with CHING K'o's and Han Kao Tsu's songs. T'ao Ch'ien's piece is simply a *fu*. The only entry here that is not a puzzle is *hsü*. There are nine Prefaces in *WH* 45-46; the real question is why HSIAO T'ung associated *hsü* with the others. He put *tz'u* and *hsü* together in *WH* 45. Perhaps the fact that half of his *hsü* are introductions to single poems led him to associate the terms. One is Shih Ch'ung's 石崇 (249-300) "Preface to the Song 'Longing to Go Home'" 思歸引序, and this may account for the word *yin* in the list.

[80] 碑 and 碣 are both epitaphs. They differ in that *pei* is inscribed on a flat stone, while *chieh* is cut on a stone column. There are no examples of *chieh* in *WH*; in ch. 58-59 there are five specimens of *pei*, of which four are epitaphs and one by Wang Ch'e 王屮 (or Chin 巾, d. 505) is an inscription commemorating the building of a temple. All consist of a prose introduction followed by a verse eulogy; see note 55.

[81] 誌 is represented in *WH* 59 by the entry 墓誌, of which JEN Fang's "Grave Inscription for the Wife of Liu Hsien" is the sole example given. It is a eulogy in four-word meter with rhyme, and formally indistinguishable from *sung* or the rhymed section of *pei*.

[82] 狀 is the same as 行狀 in *WH* 60, which gives JEN Fang's long biographical eulogy of HSIAO Tzu-liang. The form differs from *pei* in that it is all in prose, omitting any versified eulogy.

[83] 監撫: cf. *Tso chuan* (Min 2, Legge, 130): "When the ruler goes abroad [the Heir Apparent] guards the capital; and if another be appointed to guard it, he attends upon [his father]. When he attends upon him, he is called 'Soother of the Host' 撫軍; when he stays behind on guard, he is called 'Inspector of the State' 監國." (KAO Pu-ying)

[84] 縹囊緗帙, a reference to Hsün Hsü 荀勖 (d. 289), who devised the four bibliographic categories 四部 to include all books, which he stored in green bags (*Sui shu* 32.4a-b) and tied with yellow cord.

select the flowers, it is impossible, even with the best intentions, to get through the half.

Now the writings of the Duke of Chou and the works of Confucius are on a level with sun and moon, as mysterious as ghosts and spirits. They are the models of filial and respectful conduct, guides to the basic human relationships; how can they be subjected to pruning or cutting?

The works of Chuang Tzu and Lao Tzu, of Kuan Tzu and Mencius, are devoted primarily to establishing a doctrine; they are not immediately concerned with literary values. In the present anthology they too have been omitted.

When it comes to the excellent speeches of the sages and the straightforward remonstrances of loyal ministers, the fine talk of the politicians and the acuity of the sophists,[85] these are " ice melting [86] and fountain leaping,[87] gold aspect and jade echo." [88] They are what are referred to as " sitting on Mt. Chü and debating beneath the Chi Gate." [89] Chung-lien's making Ch'in's army withdraw,[90] I-chi's getting Ch'i to submit,[91] the Marquis of Liu's raising eight difficulties,[92] the Marquis of Ch'ü-ni's proposing the six

---

[85] 辯士之端 : a reference to *Han-shih wai-chuan* 7/5: " The superior man avoids the three points: he avoids the brush-point of the literary man; he avoids the spear-point of the military man; he avoids the tongue point of the sophist." (J. R. HIGH-TOWER, *Han shih wai chuan* 227).

[86] 冰釋 probably refers to *Tao te ching* 15: 渙兮若冰之將釋 " Yielding as ice as it starts to melt," used to characterize the excellent officers of antiquity 古之善爲士者. (CHANG Shao)

[87] 泉涌. There should be a less recondite source for this than the line cited by TSENG Chao: " Plans like a spring gushing " 謀如泉湧 (from the Grave Inscription for Ts'AO Ch'üan, 金石萃編 18.1a, Sao-yeh Shan-fang ed.).

[88] 金相玉振 probably alludes directly to WANG I's preface to " Li sao " (*Ch'u tz'u* 1.52b, *SPTK* ed.): " The writings of Ch'ü Yüan are truly far-reaching in their influence. . . . Of them it can be said that their aspect is of gold, their substance of jade, peerless in a hundred generations " 所謂金相玉質,百世無匹 (CHANG Shao). The 振 is perhaps contaminated by the similar (and here irrelevant) line from *Mencius* 5B/1: 金聲而玉振之 " The metal [bell] sounds and the jade [jingle] carries on."

[89] LI Shan's commentary (*WH* 42.19a) quotes a lost work, *Lu Lien-tzu* 魯連子 : " T'IEN Pa, a sophist of Ch'i, argued on Mt. Chü and debated beneath the Chi [-cheng Gate]. He defamed the Five Emperors and incriminated the Three Kings, in one day putting down a thousand opponents." 齊之辯者曰田巴,辯於徂丘而議於稷下.毀五帝,罪三王,一日而服千人. (CHANG Shao)

[90] Lu Chung-lien (GILES, *Biographical Dictionary*, No. 1408, abbreviated *BD*) dissuaded Chao from recognizing the ruler of Ch'in as emperor (as advocated by the general HSIN Yüan-yen of Wei), and the report of his indictment of CH'IN led Ch'in to withdraw its armies which were besieging Han-tan (*Chan kuo ts'e* 20.5a).

[91] LI I-chi persuaded Ch'i to join with LIU Pang in the wars that led to the founding of the Han dynasty (*Shih chi* 97).

[92] CHANG Liang, Marquis of Liu (*BD* 88), dissuaded Han Kao Tsu from re-establishing the Six Feudal States (as advocated by LI I-chi) by citing eight precedents and pointing out the differences in circumstances (*Shih chi* 55).

strategies:[93] their accomplishments were famous in their own time and their speeches have been handed down for a thousand years. But most of them are found in the records or appear incidentally in the works of the philosophers and historians. Writings of this sort are also extremely numerous, and though they have been handed down in books, they differ from belles lettres, so that I have not chosen them for this anthology.

As for histories and annals, they praise and blame right and wrong and discriminate between like and unlike. Clearly they are not the same as belles lettres. But their eulogies and essays concentrate verbal splendor, their prefaces and accounts are a succession of flowers of rhetoric; their matter derives from deep thought, and their purport places them among belles lettres. Hence I have included these with the other pieces.

From the Chou House of long ago down to this Holy Dynasty, in all it makes thirty chapters. I have named it simply the *Anthology*. The following texts are arranged by genres. Since poetry and *fu* are not homogeneous, these are further divided into categories. Within each category the sequence is chronological.

There are other points of interest in this Preface, but it is the list of literary forms which is central. This list is Hsiao T'ung's reply to the hypothetical question, " What are the forms of literature? " It is not an easy question to answer, because in any language the current labels are not all of the same order of abstraction, nor do they focus on the same criteria. Hsiao T'ung's list finds a place for thirty-eight terms, of which eleven are not represented by selections in the *Anthology*. On the other hand, ten of the thirty-seven categories in the *Anthology* have no corresponding entry in this list, and two others (Nos. 18 and 19) must be paired against different terms. This strongly suggests that Hsiao T'ung was not trying to give a complete list of all possible terms in his Preface. It also suggests that the terms there are not all of equal value, that some are merely fillers. This is borne out by the groupings, where eight symmetrical groups of four beats each are set down in an order which has no apparent significance itself and which is not followed by the contents of the *Anthology*.[94]

---

[93] CH'EN P'ing, Marquis of Ch'ü-ni (*BD* 240), became Chief Minister under Han Kao Tsu. According to SSU-MA Ch'ien (*Shih chi* 56.7a), the Six Strategies had been kept secret and he had no way of knowing what they were. However, *BD* gives a list of them.

[94] The categories in the *Anthology* correspond to the following groups in the Preface (items are numbered in order of their appearance in each list):

When we look at the *Anthology* itself, we find anomalies of the same sort. The first four entries (*fu, shih, sao, ch'i*) are no improvement over the first four in the Preface (*fu, sao, shih, sung*), since *sao* surely goes with *fu*. Whether *sung* is associated with *shih* or with *tsan* (22 and 23), depends on which criterion is decisive, "historical" derivation or subject matter. The "Sevenses" (*ch'i*) as a literary form is an example of taking a word that appears in a title and using it as a generic term. There is little in the "form" itself to justify a separate classification from *fu*, since other *fu* that are presented within a framework of dialog (e. g., "Tzu-hsü Shang-lin *fu*") are not so distinguished.

The miscellaneous official prose pieces which follow fall into three groups, 5 through 9, addressed from superior to inferior, 10 through 15, from inferior to superior, and 16-17, between

Left table:

1 賦	2 詩	3 騷	4 七		
5 詔	6 冊	7 令	8 教	9 策文	
10 表	11 上書	12 啟	13 彈事	14 牋	15 奏記
16 書	17 檄				
18 對問	19 設論				
20 辭	21 序				
22 頌	23 贊				
24 符命	25 史論	26 史述贊	27 論		
28 連珠					
29 箴	30 銘	31 誄			
32 哀					
33 碑	34 墓誌	35 行狀			
36 弔文	37 祭文				

Right table:

1 賦	2 騷	3 詩	4 頌
11 詔	12 誥	13 教	14 令
15 表	16 奏	17 牋	18 記
19 書	20 誓	21 符	22 檄
27 答客	28 指事		
31 篇	32 辭	33 引	34 序
4 頌	10 讚		
7 論			
5 箴	6 戒	8 銘	9 誄
25 悲	26 哀		
35 碑	36 碣	37 誌	38 狀
23 弔	24 祭		

equals, corresponding to three groups of four in the Preface (11-14, 15-18, 19-22). Differences here can easily be accounted for in terms of the requirement for symmetry, euphony, and rhyme in the Preface list.

No such obvious explanation appears for the next two entries (18-19) in the *Anthology*. All are dialogs, and all are essays in self-justification. The only apparent difference is that the " Reply to a Question " attributed to Sung Yü is presented as an account of something that actually occurred, where in the others the writer sets up an imaginary interlocutor to whom he addresses his reply. It is significant that in the last *WH* example, Pan Ku refers specifically to the two earlier compositions of Tung-fang So and Yang Hsiung; as with the " Sevenses " the form is created by the writer's awareness of a series of similar compositions which he consciously imitates.

There is no reason for associating *tz'u* and *hsü* (20-21), except that they seem somehow to correspond to the enigmatic group (31-34) in the Preface. *Tz'u* is actually no form at all; at least the two specimens in the *Anthology* have nothing in common except the term *tz'u* in the title of each.[95] Of the Prefaces it is interesting that no distinction is made between Prefaces written for a book by someone else (the first three are Prefaces to Classics, the *Shih, Shu,* and *Ch'un ch'iu Tso chuan*) and the introduction to one's own poem.[96]

Eulogy and Appreciation (22-23) belong together by nature and manner (both are rhymed); it is here too that one would expect the Appreciations from the Histories (26), though they do partake of the nature of Essay, with which Hsiao T'ung puts them.

The next four categories (24-27) could all have been subsumed

[95] The text of almost any verse composition can be introduced with the phrase 其辭曰.

[96] Five of the *WH* prefaces are to the author's own composition. But note that while Huang-fu Mi's Preface to Tso Ssu's " Three Capitals fu " appears with the Prefaces, Tso Ssu's own Preface is given in *WH* 4, before the text of the *fu*, but still as a separate composition. This sort of difficulty is hardly the fault of either the categories or of the compiler, but is simply one of the occupational hazards of anthology making.

under the general heading "Essays," which is the only entry in the Preface list which corresponds to any of them. The need for separate headings for 25 and 26 is questionable, since one of the "Essays from the Histories" is in fact the "Appreciation" appended to a *Han shu* biography. On the other hand, making a separate category for the "Bestowal of the Mandate" essays could be taken as a necessary gesture in the direction of legitimacy; but then why is PAN Piao's "Essay on the Royal Mandate" (*WH* 52.1a) relegated to the Essays?

The sententious maxims and exhortations going under the name of "Strung Pearls" find no counterpart in the Preface list. They appropriately precede Warning (29) which the Preface pairs with Admonition (5-6). Inscription and Dirge (30-31) go together in both lists. Lament (32) follows reasonably enough, but in the Preface list it is just as reasonably associated with Condolence and Requiem (23-24). Epitaph, Necrology, and Obituary (33-35) make a logical group in both lists; it is after them that the *Anthology* puts Condolence and Requiem (36-37), the final headings in the book.

From this survey and comparison of the two lists it is apparent that HSIAO T'ung was aware of the problems involved in a systematic arrangement of literary genres, though he did not arrive at any consistent solution. Later anthologists continued to reshuffle the terms, culminating in the thirteen classes of YAO Nai's *Ku-wen tz'u lei-tsuan* 古文辭類纂,[97] a list as influential as that of the *WH*, though still far from ideal.

---

[97] See E. D. EDWARDS, "A Classified Guide to the Thirteen Classes of Chinese Prose," *BSOAS* 12 (1948).770-88.

THE *SHIH-SHUO HSIN-YÜ* AND SIX DYNASTIES
PROSE STYLE

by
YOSHIKAWA KŌJIRŌ

Translated by
GLEN W. BAXTER

THE *SHIH-SHUO HSIN-YÜ* AND SIX DYNASTIES PROSE STYLE

BY

YOSHIKAWA KŌJIRŌ

吉川幸次郎

KYŌTO UNIVERSITY

Translated by Glen W. Baxter

[This English version, published with the author's permission, is a condensation of his " *Seisetsu shingo* no bunshō " 世說新語の文章 (" *Shih-shuo hsin-yü* and its Style "), which first appeared in *Tōhō gakuhō* 東方學報 (*The Tôhô Gakuhô. Journal of Oriental Studies*), Kyōto, 10 (1939) 2.86-109, and was reprinted with slight revisions in YOSHIKAWA's *Chūgoku sambun ron* 中國散文論 [*Studies in Chinese Prose*] (Tōkyō, 1949), pp. 66-91. Professor YOSHIKAWA's final paragraph has been transposed to the beginning of the article. Some of the material has been removed to footnotes; a few other notes have been added by the translator, and are so indicated.]

Among the manifestations of China's culture, Chinese literary style is surely one of the most distinctive. One might say that a comprehensive study of the evolution and metamorphoses of this style would constitute, in a sense, a history of Chinese literature. It is well, furthermore, to recognize the importance of style not only as a vehicle, but as a shaping factor, of philosophical attitudes and concepts. It is with these ideas in mind, rather than with the intent of making a formal linguistic study, that I have prepared this article on one phase of Chinese literary style.

The book known as *Shih-shuo hsin-yü* 世說新語 is a collection of anecdotes about officials, savants, and eccentrics who lived in the period from the last years of the Han to the close of the Chin dynasty—from the late second century to the early fifth—brought together by LIU I-ch'ing 劉義慶 (403-444), a nephew of the first Sung emperor. The stories, and the manner of their distribution under thirty-odd headings, are of value to the social historian for their reflection of attitudes and trends of thought in the period concerned.[1] However, for many centuries the book has been highly esteemed in China and Japan [2] for other reasons as well—notably for its wit, and by no means least of all for its literary style. It is the latter aspect of the book which I wish to treat, in order to consider its position and significance in the history of Chinese prose.

The style and diction of the *Shih-shuo* [3] cannot be attributed to one writer, because LIU I-ch'ing drew freely from various historical and other writings of the period. (The Liang commentary of LIU Hsiao-piao 劉孝標 includes passages which apparently served as source materials.) But if the text is a composite one, it nevertheless has a stylistic consistency which, in addition to justifying the treatment of it as a unity, suggests a common prose style for the period. In fact, comparison of this book with other works shows that its diction and usages are to a very large degree characteristic of post-Han prose down into T'ang times.

The style closest to that of the *Shih-shuo* is that found in collections of ghost stories such as the *Sou-shen chi* 搜神記 and the *I-yüan* 異苑, and the next closest that of the *Kao-seng chuan* 高僧傳 [*Biographies of Eminent Monks*]. But, in any case, the standard histories of the Northern and Southern dynasties (whether as prepared in that period or as revised under the T'ang) also are composed for the most part in a prose markedly similar to that

[1] In the same issue of *Tōhō gakuhō* (pp. 29-85) is an article on the work's historical and social background: UTSUNOMIYA Kiyoyoshi (Seikichi) 宇都宮清吉 , "*Seisetsu shingo no jidai*" ││││ の時代 ("*Shih-shuo hsin-yü* and its Era"). Cf. also Werner EICHHORN, "Zur chinesischen Kulturgeschichte des 3. und 4. Jahrhunderts," ZDMG 91 (1937) .2.451-483.—Tr.

[2] It is noted in the Heian *Genzai shomoku* 見在書目, and a manuscript copy from that period is still extant. Tokugawa Confucianists, in particular ŌGYŪ Sorai, regarded it highly; they produced no less than ten commentaries on it.

[3] This abbreviation, used hereafter for convenience, may have been the original title of the book. It is so listed in the *Sui History* and *Old T'ang History*; it appears in the *New T'ang History* as *Liu I-ch'ing shih-shuo* [*Liu I-ch'ing's Anecdotes*], perhaps to distinguish it from an earlier *Shih-shuo* now lost. It was also referred to in T'ang times as *Shih-shuo hsin-shu* │││書 [*Anecdotes Newly Written*]; Kyōto University has published a photographic reprint of a T'ang manuscript copy so entitled. According to KONDO Moku 近藤杢 , *Shina gakugei daiju* 支那藝學大辭彙 (Tōkyō, 1940), p. 661, the "Bibliography" of the *Sung shih* contains the earliest known reference to the book as *Shih-shuo hsin-yü*. The title may be rendered succinctly, if not word for word, as *New Anecdotes*.—Tr.

of the *New Anecdotes*. This study, then, is not made for the sake of a single text, but because it seems to me that the text under discussion most clearly exhibits the characteristics of narrative and expository composition during the period loosely referred to as that of the Six Dynasties.[4]

To illustrate some of the characteristics of these writings, several excerpts from the *Shih-shuo* are reproduced in the course of this article. References are to the *Ssu-pu ts'ung-k'an* edition.[5]

Excerpt I (1A.4b)

華歆王朗俱乘船避難．有一人欲依附；歆輒難之．朗曰『幸尚寬，何爲不可？』後賊追至；王欲舍所攜人．歆曰『本所以疑，正爲此耳．既已納其自託，寧可以急相棄邪？』遂攜拯如初．世以此定華王之優劣．

HUA Hsin and WANG Lang together were in a boat fleeing the disorders. There was a person who wanted to attach himself to them; Hsin, for his part, objected to this. Lang said, "It happens there is still room, why isn't it all right?" Later brigands were overtaking them; WANG wanted to get rid of the man they had brought along. Hsin said, "This is just why I hesitated in the first place. Since we have already acceeded to his request, can we throw him out on account of an emergency?" So they took their protégé along as before. Society from this has judged the superiority and inferiority of HUA and WANG.

Note in the above passage the large proportion of characters (those underlined) which were grammatically unessential, the elimination of which would still leave a series of complete and intelligible constructions. These characters represent what the Chinese call *chu-tzu* 助字, "aid-words," since (at least in the sense in which they are used above) they have neither substantive, adjectival, nor verbal force, functioning only to supplement or "aid" the essential characters or *shih-tzu* 實字. To designate these auxiliary or aspective words the often-used term "particles" may be adopted for convenience, though of course they are not particles in the sense used for polysyllabic inflected and agglutinative languages.

In written Chinese it is always possible to form a complete and

---

[4] The belles-lettres of the period, as represented in the *Wen-hsüan*, are less easy to treat in this connection without lengthy qualifications, but even this type of writing shares with the *New Anecdotes* general tendencies of diction and usage.

[5] Western style punctuation, and underlines for discussion purposes, have been added by the translator.

intelligible construction from *shih-tzu* alone, the order of the characters sufficiently indicating their relations. If all the so-called particles are removed from Excerpt I, its basic meaning is still clear. Even without 俱 " together " it is obvious that HUA Hsin and WANG Lang both fled. Omit 有, and instead of " there was a person who wanted to attach himself " we have simply " a person wanted to attach himself," but " there was " is understood. Any one reading the story would also understand that the new-comer " wanted " (or " started ") to join them, with or without the character 欲. The character 輒 " for his part "[6] is not needed to set off the subject " HUA Hsin," nor is 之 " to this " needed to complete the verb " objected."

WANG Lang's statement, " It happens there is still room," actually could be reduced to the single word 寬 " room " (or " space "). Such extreme terseness is by no means unknown in Chinese, and the ideas conveyed by the two preceding characters would not be lacking in the reader's mind. In " why isn't it all right," 為 could easily be omitted; logically the clause could be reduced to the single character 可.

Since it is obvious that the pursuit by bandits took place " later," 後 is unnecessary. Where WANG wants to get rid of the extra man, the omission of 欲 might obscure the meaning temporarily, but certainly is grammatically possible. The first part of HUA's protest could be expressed adequately by various shorter combinations, such as 本疑為此耳 ; it would even be possible for 疑此 to stand alone. As for " Since we have already acceded to his request," either 既 or 已 alone would explicitly place the action in the past, and neither is really necessary because the reader knows from the preceding text that the request had been granted in the past. Similarly the idea here rendered as " his " is redundantly carried by two characters, 其 and 自, both of which could be deleted without causing any misunderstanding as to whose request was granted. The sequel, " can we throw him out on account of an emergency? " is of course a leading question implying a nega-

---

[6] LIU Ch'i 劉淇, *Chu-tzu pien-lüeh* 助字辨畧 5, *s.v.* 輒 (p. 282, K'ai-ming edition, Shanghai, 1940), quotes this passage and explains the character as meaning 獨 or 特.—Tr.

tive answer; either the initial 寧可 or the final 耶 would be sufficient to give the question such a turn, nor would it be difficult,
given the context, so to understand it in the absence of all three
characters. Again, there is no real necessity for the presence of
以 to show the causal relation between 急 " emergency " and 棄
" throw [him] out." Most obviously dispensable of all is the
character 相 , which here has almost no meaning. (Its slight
nuance will be discussed later.)

In " So they took their protégé along as before," the initial 遂
could easily be deleted. Neither 如 nor 初 could very well be
omitted singly, but both together could be dispensed with, since
they add nothing to the meaning that is not already obvious.

In the final sentence of Excerpt I, elimination of 以 and 之
would not alter the sense, which then might be rendered " Society
here judged HUA and WANG superior and inferior." Nor is 此
" here " essential.

Almost half the characters in this passage are " particles "
which could be left out or avoided. Of course the use of such
" particles " was nothing new; certain of them are found in the
earliest Chinese prose and poetry, sometimes to convey shades of
meaning, sometimes for euphony, sometimes for both purposes.
They appear increasingly in the ancient philosophical writings,
and in such works as *Chan-kuo ts'e* and *Tso chuan*. But in no
text prior to the period of *Shih-shuo hsin-yü* are aid-words employed in such number. If SSU-MA Ch'ien or PAN Ku had written
the anecdote told in Excerpt I, the proportion of non-essential
to essential words certainly would have been much smaller.

In *Shih chi* and *Han shu* the connections between words or
phrases, and the causal or other relations between actions or
conditions, ordinarily are left for the reader to infer from the
order of the essential words. This terseness is admired as " classic,"
but compared with the style of *Shih-shuo* the prose of these histories is almost primitive. It is like a simple outline drawing;
however complex the texture and tension implicit within the
outline, there is no modeling or chiaroscuro to show clearly the
relation of one surface to another. In *Shih-shuo* such modeling is
effected through the liberal use of particles. As a result the prose

becomes more supple rhythmically, and also more subtle in what it is able to convey. It comes nearer to the rhythms and modulations of speech, with the attendant advantages and disadvantages of greater prolixity. As with speech, its subtleties sometimes become vagueness.

Normal speech in any language seldom approaches the density of formal literary composition, which in everyday conversation would be difficult for both mind and tongue. Both difficulties are intensified in the case of monosyllabic Chinese. Apart from the problem of homophones, a rapid succession of independent syllables all charged with definite ideas would be psychologically and physically very taxing. Relaxation and rhythmic flow are achieved in spoken Chinese by the liberal use of particles and by polysyllabic (usually disyllabic) compounds. This is perhaps putting the cart before the horse. It is more logical to conjecture that the Chinese always tended to speak in what, for practical purposes, we may term polysyllables; but that when they began to write, the ideographic nature of their script made it unnecessary to represent by two syllables what could be conveyed to the eye by one, and that the nature of their early writing materials made it undesirable to do so. However, eventually free use of compounds and of particles crept into writing as well. The *Shih-shuo hsin-yü* is strongly characterized by both.

The stylistic effect of compounds will be dealt with further on. As for particles, the *Shih-shuo* not only uses them more frequently than does the prose of earlier periods, but uses new ones, and old ones in new ways. One example is 相 in " can we throw him out on account of an emergency? " In Ch'in and Han writings *hsiang* is used only in the sense of " mutually " or " together with," being limited to cases where the action is reciprocal or equal on both sides. In the present instance *hsiang* has no very definite or important meaning, merely accompanying the verb 棄 for emphasis of a sort and giving the speech a looser texture. The *Shih-shuo* sometimes uses *hsiang* in the older, more " correct " sense, but cases of unilateral action such as this *hsiang-ch'i* are more numerous. Such usages of *hsiang* have indeed some relation to its stronger sense, in that they are still limited to actions of which the originator shows consciousness of the feelings or fate of the other party. The new usage probably resulted from overuse of *hsiang* due to the tendency toward two-syllable combinations.

In modern Chinese the use of *hsiang* in a unilateral sense has all but disappeared, surviving in a few expressions such as 相信 (to believe or trust, whether or not the trust is mutual) and 相思 (to love, whether or not the love is requited) .[7]

Similar to the vague uses of *hsiang* are the cases of *pien* 便 and *fu* 復 as they appear in

Excerpt II (IB.25ab)

支道林許謝盛德共集王家．謝顧謂諸人，『今日可謂彥會．時旣不可留．此集固亦難常．當共言詠，以寫其懷．』許便問，『主人有莊子不？』正得漁父一篇．謝看題，便各使四坐通．支道林先通，作七百許語．叙致精麗，才藻奇拔．衆咸稱善．於是四坐各言懷．言畢，謝問曰，『卿等盡不？』皆曰『今日之言，少不自竭．』謝後麤難，因自叙其意，作萬餘語．才峯秀逸，旣自難干，加意氣擬託，蕭然自得．四坐莫不厭心．支道林謂謝曰，『君一往奔詣，故復自佳耳．』

CHIH Tao-lin (CHIH Tun 遁), Hsü [Hsün], HSIEH [An], and [WANG] Sheng-te (WANG Meng 王濛) were gathered together at WANG's house. HSIEH looked around and said to the group, "Today's is quite an eminent assembly! The occasion once past will be beyond recall. It's certainly hard to regard this gathering as routine. It would be well to intone together and thus express our thoughts." Hsü then asked, "Has our host the *Chuang tzu?*" and happened to turn to the "Fisherman" passage. HSIEH glanced at the theme, and then each of the party was required to develop it. CHIH Tao-lin developed it first, composing more than seven hundred words. His exposition was quintessential, his brilliance outstanding; the group as a whole praised his excellence. Thereafter each of the others present put his thoughts into words. When the speeches were over, HSIEH said, "Is everybody finished?" They all said, "Today's talk leaves little unexhausted." HSIEH after this summarily refuted them, and then related his own ideas, composing over ten thousand words. [As for the content,] his talent was of an eminence scarcely to be touched, and [as for the delivery,] his emphasis was so rightly placed as to

[7] Dr. YAMADA Yoshio 山田孝雄 has pointed out that the much-used *sōrōbun* expressions *aimoyōshitaku* 相催度, *ainari sōrō* 相成候, and *aihikae mōshi sōrō* 相控へ申候 originated in translations of Six Dynasties prose. Cf. his *Kambun no kundoku ni yori tsutaeraretaru gohō* 漢文の訓讀により傳へられたる語法 [*Expressions Handed Down through Japanese Translations from Chinese*] (Tōkyō, 1935), 196-201.

give complete satisfaction. None of the company failed to be impressed. Chih Tao-lin said to Hsieh, " You went straight to the point—it was really superb."

In Ch'in and Han prose *pien* usually, if not always, has the significance of " instantly; " in the *New Anecdotes* it often serves only as a kind of connective, with hardly more force than " and " or " so "—something like the Japanese *de*, or the light use of *chiu* 就 in modern Chinese. In earlier writings *fu* always carries its essential meaning " once again," but in *Shih-shuo* it often has almost no significance, serving merely to keep the prose from being too staccato.

Such shifts of usage from one period to another, or from one style to another, were not taken into account by Japanese *kambun* scholars. *Pien* was always given the *kun* reading *sunawachi* (in the sense of " thereupon "), the same as 卽, 則, 乃, etc., while *fu* (like 又 and 亦) was invariably read *mata,* " again." Teachers may have been aware of the differences, but their *kun* readings must inevitably have caused misunderstandings on the part of students with regard to texts like the *Shih-shuo*.

A word which assumes a new function in Six Dynasties prose is 是, as in

Excerpt III (2B.15b)

林公云,『王敬仁是超悟人.』

Master Lin (a Taoist savant) said, " Wang Ching-jen is a man of superior understanding."

Shih here obviously does not mean " this " or " this one " but acts as a copula precisely as in modern Chinese, e. g., " *Wo shih Chung-kuo jen.*"

In early Chinese, direct juxtaposition of subject and predicate was enough to show that the two were one. In Ch'in-Han writings *shih* is often used to set off an entire statement as grammatical subject, but there are no more than three or four instances to my knowledge where it acts for a simple sentence-subject, as is common in the *Shih-shuo*. As in the case of the particles already discussed, the new usage to some extent robbed the character *shih* of its original meaning, but this time the result was the opposite

173

of vagueness. If in earlier Chinese direct juxtaposition of subject and predicate indicated that the two were one, the development of a copula made the idea unmistakable.

The new modes of expression described above appear to have developed from careless overuse of certain words until their meanings dwindled or changed. There are other new usages in the *Shih-shuo* which are difficult to derive from any meaning of the same characters in earlier writings. One of these is the frequent appearance of 都 to reinforce a negative, as in

Excerpt IV (1B.15a)

衞玠始渡江,見王大將軍,因夜坐.大將軍命謝幼輿.玠見謝,甚說之, 都不復顧王,遂達旦微言.

Only after WEI Chieh had crossed the River did he meet up with Generalissimo WANG; they then had a night visit. The Generalissimo summoned HSIEH Yu-yü. When WEI Chieh met HSIEH he was delighted, and paid no attention at all to WANG. And until dawn there was subtle talk.

The use of *tu* (or *tou* in modern Peking colloquial) in "paid no attention at all to WANG" is still idiomatic in modern Chinese, fully as much so as that of *ping* in 並不 or 並沒有, "not at all" or "absolutely none." It does not correspond to any use of *tu* in Ch'in-Han texts, where in fact any kind of particle emphasizing a negative is rare. Since a negative is absolute by definition, a terse writer feels no need to reinforce it; not until the more leisurely pace of the *Shih-shuo* came into fashion did such combinations appear.[8]

Another expression for which it is hard to find any antecedent is 定是 in

Excerpt V (3A.45ab)

襄陽羅友有大韻...在益州語兒云,『我有五百人食器.』家中大驚其 由來清而忽有此物.定是二百五十㭊烏㯻.

Lo Yu of Hsiang-yang had a grand manner ... In I-chou he told his children, "I own a dining service for five hundred persons." His household was greatly startled that he, who had always been poor, should possess such

---

[8] *Chūgoku sambun ron* includes a chapter (pp. 142-149) on the emphatic negative in Chinese.

a thing. Actually it turned out to be two hundred fifty black [divided] cakeboxes.[9]

Ting seldom appears as a " particle " in the *Shih chi* or *Han shu*, and from T'ang times on, when used adverbially it either meant " definitely " or implied a conviction of certainty, something like " I'll bet . . ." But to render *ting* in the above passage as *sada-mete*, the reading followed in Tokugawa editions of the *Shih-shuo*, is to miss or at least misrepresent the sense. *Ting-shih* here is a mode of anticipating surprise at something unexpected, like the modern colloquial *kan shih* 敢是. There is nothing to which it corresponds precisely in Ch'in or Han texts.

Apart from frequent two-beat rhythmic units formed by the combination of particles or by attaching them to basic words, the *Shih-shuo* constantly uses redundant two-syllable compounds where a single character would answer the purpose as far as essential meaning is concerned. Examples are 依附, 既巳, 詠言, 叙致, 精麗, 才峯, 奇拔, 才藻, 秀逸, 意氣, 擬託, and 奔詣. Such combinations are by no means unknown in Ch'in-Han texts, but there they are far less frequent. Together with expressions involving unprecedented uses of particles, these compounds add new dimensions to written communication in Chinese.

Where did writers get these new turns of phrase? Their similarity to modern colloquial expressions suggests that they came into literature from the spoken language of those times. It appears that the prose of the *New Anecdotes* reflects a progressive trend of the Chinese tongue toward greater minuteness and detail through increasing use of particles and the coining of two-syllable compounds, the two factors abetting one another. How far the trend has proceeded today can be illustrated by retelling part of Excerpt I in modern *pai-hua*:

華歆和王朗一塊兒乘着船去避難 . . . 有一個人要來依附他們 . . . 後來賊人追到了；王就要捨棄他們所攜帶的人 . . .

Of course, it cannot be proved that the spoken language is much

[9] For explanation of the obscure term 沓烏楳 cf. *ibid.*, p. 97, where the passage is further dealt with in a chapter (97-141) on the uses of 定, 將無, 將不, 將非, 頗 and 何物 in various Six Dynasties texts.—Tr.

175

further elaborated in these directions than was the colloquial of
the fourth and fifth centuries. Nobody knows exactly how the
story might have been told then by word of mouth. Unlike
modern *pai-hua* authors, those who wrote the *Shih-shuo* and
similar texts of its period did not attempt to adopt bodily the
patterns of speech. But they do seem to have been influenced by
them to a greater extent than earlier literati.

This development is of great significance in the history of Chinese prose. In China there had always been—and there continued
to be upheld—the idea that speech and written composition were
vastly different things, and that never the twain should meet.
Canonical sayings attributed to Confucius express this idea in
converse ways: " Writing does not use up all that is in words,"
and " When words are unpatterned, their effectiveness is not
far-reaching." [10]

In a sense these statements obviously are true for all languages.
An extreme interpretation of them, divorcing literary composition
from the development of the Chinese language, eventually lent
support to a return to ancient styles as models. In the eighth
century HAN Yü and LIU Tsung-yüan, reacting against certain
excesses of elaboration which finally vitiated Six Dynasties and
T'ang prose, propagated their neo-classic *ku-wen*, in which they
imitated the terseness and directness of Ch'in and Han writings.
Though their movement did not immediately achieve a lasting
success, it re-emerged in Sung times and thereafter was regarded
as the mainstream of prose, being practiced down to the literary
revolution of the twentieth century and in some cases even later.
Ku-wen also had great influence in Japan; the *kambun* compositions of the Tokugawa Confucianists were all in this style (despite
their admiration for the *Shih-shuo*), and *kambun* lessons taught
in Japanese schools today are still constructed in imitation of
Ch'in-Han prose.

[10] Cf. LEGGE, (1) *The Yî King* [= *SBE* XVI], 376-377: " The Master said:—
' The written characters are not the full exponent of speech, and speech is not the full
expression of ideas . . .' " and (2) *The Ch'un Ts'ew; with the Tso Chuen* [= *The
Chinese Classics* V], Pt. 2, p. 517: " ' Chung ne said, " An ancient book says, ' Words
are to give adequate expression to one's ideas; and composition, to give adequate
power to the words.' Without words, who would know one's thoughts; without
elegant composition of the words, they will not go far." . . .' "—Tr.

The *ku-wen* movement largely suppressed the influence of contemporary idiom on formal prose. In fairness, however, it must be remembered that HAN Yü and his colleagues were reacting primarily not against " naturalness " in composition, but against the abuses of excessive complication, elaboration, and ornamentation to which increased vocabulary and new modes of expression were put in the later Six Dynasties and the first two centuries of the T'ang. These abuses had not yet gone to extremes in the era of the *Shih-shuo*, in which elaborate parallelism, for instance, hardly figures; but this text does show signs of the excrescences which later overloaded and weakened the medieval prose structure.

It is not a simple matter to separate these symptoms categorically from signs of strength. On the one hand, the ubiquity of particles and other " extra " words gives the prose of the *Shih-shuo* a suppleness enabling it to encompass nuances of thought and expression only hinted at in earlier styles, whether those styles be called classical or primitive. On the other hand this elasticity sometimes verges on limpness, probably exceeding the degree of convolution demanded by everyday speech. As a case in point, it is hard to see any real necessity for the involved constructions of WANG Lang's last speech in Excerpt I. Indeed, even in diffuse modern *pai-hua* one would be hard put to it to duplicate precisely all the twists of those four clauses. There is nothing today to correspond to *tzu* in 自託, and one suspects that there was nothing like it in LIU I-ch'ing's day either. Such use of *tzu* may have been a purely literary invention. It is by no means the only particle in the *Shih-shuo* which gives the impression of artifice. In many cases words of this sort hardly serve the purpose of rhythmic fluency, since in effect they impede the prosodic flow rather than facilitate it. Their presence is often hard to explain except on the basis of a predilection for embellishment.

The impression is inescapable that the *Shih-shuo* uses particles and compounds to a large extent for ornamental purposes, in contrast to earlier texts such as the *Shih chi* and the *Han shu* where they are used almost always for the practical purpose of clarifying the meaning where necessary. True, the *Shih-shuo* text

177

often succeeds in expressing its meaning in finer gradation, but sometimes its particles add nuances so delicate as to be either pointless or puzzling. Now and then—to the modern reader at least—they actually obscure the meaning instead of clarifying it.

As observed earlier, the abundance of particles and the vogue for compounds were mutually propagative. In addition to the coupling of words meaning virtually the same thing, as in 既已, two-syllable units were constantly formed either by attaching an unnecessary particle to an essential word, or by using the particles themselves in pairs. *Fu* and *shih* appear in combination not only with each other (復是), but with countless other words varying from essential to supererogatory: 故是, 迺是, 直是, 自是, 並是, 正是; and 不復, 非復, 勿復, 無復, 亦復, 乃復, 已復, 雖復, 時復, 豈復.

What causes brought about the overabundance of "extra" words in the *Shih-shuo* text? It has already been posed that sensitivity to a natural trend of the Chinese language was partly responsible. But the special character of this prose cannot be understood without considering certain other factors.

One of these is the wide vogue of philosophical disquisition in the post-Han period. That era was one of those in which the Chinese philosophical spirit was most unconstrained by precedent or authority. Political power was so parcelled out that it was to be centuries before a central power like that of the Han would again be achieved. As a result, scholars could not effectively be held to any line of orthodoxy, and were able to indulge in free speculation. It became the fashion to hold arguments and long involved conversations like those referred to in Excerpts II and IV. These were the so-called *ch'ing-t'an* 清談 or *ch'ing-yen* 清言 of Wei and Chin times. The chief texts used as points of departure for these divigations were the *Lao tzu*, *Chuang tzu*, and *I ching*. Participants in such sessions no doubt tried to present their ideas and intuitions as fully and minutely as possible. For that reason the abundance of particles in *ch'ing-t'an* sessions was probably greater than in everyday speech of the period.[11] One plausible

---

[11] Among the papers of Hsi K'ang 嵇康 (223-262) in *Hsi Chung-san chi* 嵇中散集 are the only surviving texts which appear to be direct transcriptions of con-

explanation for the unprecedented number of particles in the *Shih-shuo* is that the text reflects *ch'ing-t'an* diction, especially in the notably frequent use of words with the flavor of 自，本，正，and 亦. The new usage of 是 and emphatic expressions like 都不 may well have passed into literature from *ch'ing-t'an*.

Significant here is not merely the permeation of the text by modes of expression characteristic of *ch'ing-t'an*, but the impulse to make language itself a component, rather than simply a carrier, of philosophy. There is a difference between describing or explaining philosophical theory in words, and seeking to embody it in subtle turns of phrase or fragmentary utterances, as does the *Kung-yang Commentary* to the *Springs and Autumns*. The *ch'ing-t'an* conversationalists sometimes went to great lengths in expounding their theories (as indicated in Excerpt II), but on the other hand they often sought to embody basic ideas in the subtle wording of a brief question or reply.

An example is the *Shih-shuo* passage [12] in which JUAN Hsüan-tzu 阮宣子 (JUAN Hsiu 修), being asked by WANG I-fu 王夷甫 (WANG Yen 衍) the differences and similarities between Taoism and Confucianism, replies " Chiang wu t'ung " 將無同—"Aren't they the same? " This " three-word commentary " promptly became famous for its terseness and subtlety, but if it was readily understood in *ch'ing-t'an* circles, by Southern Sung times (after the victory of *ku-wen*) different people interpreted it in diametrically opposed ways—not only because of their varying philosophical preferences but because of their uncertainty as to the exact force of the old colloquial expression *chiang-wu*.[13] Had

---

versations in this period. However, there are passages in the commentary by KUO Hsiang 郭象 to the *Chuang tzu* and in that of CHANG Chan 張湛 to the *Lieh tzu* which appear to be derived from such transcriptions.

[12] IB.14a-b. Almost the same passage appears in *Chin shu* 49 (Biography of JUAN Chi 籍), where the speakers are JUAN Chan 瞻 (JUAN Hsiu's nephew) and WANG Jung 戎, of whom WANG Yen was an admirer. The *Po-na* edition (*ts'e* 11, *lieh chuan* 19.3b) has 无 instead of 無.—Tr.

[13] This uncertainty has persisted. Lien-sheng YANG, in a review of the FUNG-BODDE *A History of Chinese Philosophy* (*HJAS* 17 [1954]2.478-483) thinks it likely that FUNG Yu-lan did not quite catch the nuance. YANG writes (pp. 480-481):

Now the three characters which constituted the famous reply (translated " Can they be without similarity? ") are *chiang-wu t'ung* . . . The compound *chiang-wu* is an old colloquial expression

JUAN Hsiu wished to state unequivocally that Confucianism and Taoism were the same, he could have said simply "Tʻung." Had he wished to state that they were categorically different, he could have replied "Wu tʻung." But he sought in three words to express something above and beyond either, or perhaps, at least, to give an air of enlightened scepticism toward doctrines. A conscious desire for such refinement of thought and language is evident all through the *Shih-shuo*.

The other important factor in the overuse of particles and compounds is a formal one: the development of the four-six style. As evidenced by the accompanying excerpts, four-character and six-character clauses are already very frequent in the *Shih-shuo*. To a certain extent this tendency is as natural as the tendency to two-syllable compounds; two pairs make four, three pairs make six, and more than three pairs becomes unwieldy. A judicious admixture of four- and six-character sentences is rhythmically pleasing. But page after page of them, with scarcely ever any irregularity, is pushing a natural tendency to the point of monotony, and of artificiality as well. At that point writers begin padding. In the time of LIU I-chʻing and his sources, convenient padding materials were newly to hand in the wealth of particles used in *chʻing-tʻan*—words which would not change the essential meaning of a shorter phrase, but which in addition to filling it out to the required number of four or six, would add enough nuance to suggest a philosophic attitude. This is not to say that genuine philosophic attitude was absent, or that particles were used in the *Shih-shuo* primarily for padding; but the impression of superfluity is sometimes almost as strong as the impression of subtlety.

Note that the speech of WANG Lang against getting rid of the extra man, in Excerpt I, is in the four-six form (except for the final 耶, which has the effect of a kind of tailpiece or "end

used to ask a rhetorical question implying the mild suggestion "Wouldn't you agree?"... Its meaning and flavor have been discussed by Chinese and Japanese scholars from Sung times on. For a summary of such discussions, see YOSHIKAWA Kōjirō ... *Chugoku sambun ron* ..., pp. 107-120, where he translates the reply as 同じのではありますまいか. The main purpose of the reply was to underline the idea of similarity and to echo the suggestion that the difference between Confucianism and Taoism was only superficial. Professor FUNG's interpretation is different in his taking the reply to be a yes-and-no answer. —Tr.

quote "). Though some of the particles might well have been used in an actual speech of this sort, their selection and placement here seem designed to serve the purpose of filling out two pairs of four characters and two of six.

The literary aesthetic of the *Shih-shuo* writers is articulated, as well as exemplified, in the deliberate word-music (note that the speech is all in sixes) of

Excerpt VI (IA.46ab)

道壹道人好整飾音辭．從都下還東山，經吳中，已而會雪下．未甚寒．諸
道人問在道所經．壹公曰，『風霜固所不論．乃先集其慘澹．郊邑正自飄
瞥，林岫便已皓然．』

The monk Tao-i had a taste for polished diction and euphonious expressions. He left the capital to return to Tung-shan (Chekiang), and passing through Wu ran into a snowfall. It was not extremely cold. The monks asked what befell him on the way. Master [Tao-]i said, "The wind and frost, certainly, were nothing to speak of. However, as a prelude, [the sky] amassed its gloom. No sooner had the countryside begun twinkling [with snow] before my eyes, than the forested hills were already white."

In one sense the luxuriance of Six Dynasties prose style shows a reaction against the starkness of Later Han prose style, which is characterized by a kind of neo-classicism as compared with that of the Former Han. To use as illustration the outstanding representative works of the two Han periods, the naturalness of the *Shih chi* was subjected to some constraint in those portions adapted from it for the *Han shu*. An example may be given which is directly related to the present study. Ssu-MA Ch'ien occasionally, though rarely, followed a monosyllable subject with the character 是 : e. g., Yüan Ku 轅固 answers Empress-Dowager Tou 竇 (on the subject of the *Lao tzu*) with the words 此是家人言耳: " These are words for a slave." [14] Now Pan Ku's version of this passage omits *shih* and replaces the final *erh* with a more literary

[14] *Shih chi* 121.7b (*Po-na* ed., *ts'e* 29). Commentators differ as to the precise meaning of *chia-jen* here; Yoshikawa gives the passage only in Chinese, but in his *Kan no Butei* 漢の武帝 [= *Iwanami shinsho* 岩波新書 24, Tōkyō, 1952], p. 17, he paraphrases: あんなものは，奴隷／讀む本でございます． (For a discussion of this and other works of the author, cf. Burton Watson, " Some New Japanese Translations of Chinese Literature," *FEQ* 14.2 [February 1955].245-249.) —Tr.

particle: 此家人言矣 .[15] It is logical to suppose that the modes of
expression which later appeared in the *New Anecdotes* were already flourishing in the colloquial of Later Han times, but if so,
PAN Ku admitted none of them whatsoever.

In another respect, however, the *Shih-shuo* represents rather a
continuation of Later Han literary tendencies: the four-character
unit first became frequent in the *Han shu*, and together with its
extension, the six-character clause, it became standard in the
Shih-shuo and other works of its period. By adopting on the
one hand the symmetrical construction favored by PAN Ku, and
on the other hand semi-colloquial elements which he rejected, and
by pushing both to extremes, the authors of these writings created
a new style quite foreign to the classics. Its luxuriance remained
the almost uncontested stylistic criterion of Chinese prose for half
a millenium, and its genuine riches have continued to compel
admiration.

However, its devices of diction and form were dangerously
susceptible to abuse. The indiscriminate use of particles (that is,
particles as defined in this study) eventually led to over-refinement
and frequently to obscurity, while the ultimate crystallization of
the four-six vehicle negated the rhythmic elasticity which gave
early post-Han prose much of its freshness. And the constant
parallelism which became a concomitant of this form in the late
Six Dynasties virtually eliminated all spontaneity, reducing composition to a formula.

Thus a style at first distinguished for pliancy and expressiveness
degenerated into the stultifying artificiality and preciosity against
which HAN Yü and LIU Tsung-yüan rebelled. Although the initial
success of their *ku-wen* movement was short-lived, it reemerged
in Sung times to overthrow the " new " style which had outlived
both its newness and its effectiveness.

It is ironic that just as Chinese history in other respects was
entering " modern " times, formal prose reverted to imitation of
the ancients, one kind of artificiality replacing another. However,
ku-wen not only had the prestige of association with the Classics;
it had the practical advantage of being easy to imitate and hard

---

[15] *Han shu* 88.19a (*Po-na* ed., *ts'e* 26).

to distort. If at its best it lacked the shimmering texture which Six Dynasties prose sometimes had, at its worst it lacked the tangled intricacies which too often made that texture all but impenetrable.

In the matter of long-term comprehensibility, the neo-classic style profited by its independence of contemporary modes of expression, which in a sense placed it outside time. At any rate it remained the standard form of Chinese prose from Sung times down to the literary revolution of the twentieth century.

This did not prevent the best representatives of Six Dynasties prose from being esteemed in retrospect, and the style of the *Shih-shuo hsin-yü* in particular has never ceased to have its admirers.

METRICAL ORIGINS OF THE *TZ'U*

by
GLEN W. BAXTER

METRICAL ORIGINS OF THE *TZ'U*

Glen William Baxter

[I am deeply indebted to Professor James R. Hightower, Mr. Achilles Fang, and Professor Lien-Sheng Yang for guidance, assistance, and corrections in the preparation of this study.]

As a literary genre, the Chinese song-form known as *tz'u* 詞 reached its widest popularity and its most varied practice in the Sung dynasty; readers today associate it chiefly with that period. However, as is suggested by the title of the first extensive *tz'u* anthology, the tenth-century *Hua-chien chi* 花間集,[1] the species was already in flower during the Five Dynasties. To pursue the figure, its taproots reached well into the T'ang period; by the middle of the ninth century its early shoots were sufficiently well developed to be distinguishable from the *shih* 詩, of which it is generally considered a mutation.

The *tz'u* has been defined as " a song-form characterized by lines of unequal length, prescribed rhyme and tonal sequence, occurring in a large number of variant patterns, each of which bears the name of a musical air."[2] It would be hard to phrase a convenient definition more aptly, but before concentrating on the matter of metric it will be useful to comment on or qualify each element.

In origin the *tz'u* is certainly a song-form, and it remained so during the centuries of its greatest popularity and significance. When it began to be replaced by new song-forms in late Sung and early Yüan times, it became a vehicle for antiquarian literary artifice, and except for isolated instances, thereafter it was a song-form only by courtesy.

It is true that most *tz'u* are in lines of unequal length, but there

---

[1] Collected by Chao Ch'ung-tso 趙崇祚 (Preface dated 940), modern editions: Hua Lien-p'u 華連圃, *Hua-chien chi chu* ｜｜｜注 (Shanghai: Commercial Press, 1935); Li Ping-jo 李冰若, *Hua-chien chi p'ing-chu* ｜｜｜評注 (Shanghai: K'ai-ming Book Company, 1935).

[2] J. R. Hightower, *Topics in Chinese Literature* (Cambridge, 1950) 80.

are numerous exceptions, as will be shown by a glance at the first *chüan* of the *Tz'u-p'u* 詞譜 or the *Tz'u-lü* | 律.[3] Those *tz'u* in regular four-line stanzas are nearly all by T'ang and Five Dynasties writers, and are significant in the ensuing discussion of the origin and development of the form.

It is undeniable that in Sung times some, and later all, of the *tz'u* writers who copied earlier patterns instead of inventing new ones, bound themselves to the sequences of tones and rhymes used by their predecessors. But those predecessors themselves by no means seem to have been so meticulous.[4] Naturally those writing to the same tune put their end rhymes, at least, in the same places; but where *A* used rhymes in the level tone, *B* might use rhymes in a deflected tone. The two might differ quite freely in their sequence of tones within lines. A third poet might write in agreement with *A* as to rhyme, and with *B* or with neither as to tonal sequence.[5] However, each of the three specimens could be used as a model by some later poet when the tune itself, being lost, could no longer serve as guide, and when he would simply " fill

---

[3] Compendia of *tz'u* patterns. For a discussion of these and similar reference works, see my " A Bibliographical Note on the *Ch'in-ting tz'u-p'u*," *HJAS* 14 (1952) .668-671. The term *tz'u-p'u* is used generically by bibliographers to cover all such works.

Wherever possible, the *Ssu-pu ts'ung-k'an* (abbreviated *SPTK*) has been used for references to Chinese literary texts, and the *Po-na-pen* (abbreviated *PNP*) for references to the Standard Histories. In addition, I have used the Chung-hua Book Company's typeset series, *Ssu-pu pei-yao* (abbreviated *SPPY*).

For economy of reference, wherever possible without involving significant textual variation, numerous works are cited from *Yüeh-fu shih-chi* 樂府詩集 (*SPTK* ed.) (abbreviated *YFSC*), an anthology of real and imitation song-words compiled by Kuo Mao-ch'ien 郭茂倩 (Sung); and from the officially compiled *Ch'üan T'ang shih* 全唐詩 (Preface, 1703), in the T'ung-wen shu-chü 同文書局 edition.

Several short T'ang prose works are cited from *T'ang-tai ts'ung-shu* 唐代叢書 (abbreviated *TTTS*), a collection of uncertain provenience which took its present form in the eighteenth century. The relevant passages have been checked with other texts. This *ts'ung-shu* was chosen partly for its inclusiveness and partly because readers may find some information on all its component works in E. D. EDWARDS' *Chinese Prose Literature of the T'ang Period* (2 vols., London, 1937), which is based on it. The edition I cite is that of the Chin-chang t'u-shu-chü 錦章圖書局 (Shanghai, 1929) in twelve *ts'e* 冊, in each of which the folios are numbered consecutively.

[4] See note 60.

[5] For illustrative versions of " Ting-hsi-fan " 定西番 by three different authors, see *Tz'u-p'u* 2.28b-30b.

187

in a *tz'u* " (*t'ien-tz'u* 填 |) , following a specimen at hand.[6] Accordingly the various *tz'u-p'u* reproduce a specimen after each model as a separate form (*t'i* 體) of each pattern (*tiao* 調); if any poet has added to or subtracted from the number of characters used, yet another form is considered to have been established.

For these reasons Wu Mei-sun 吳眉孫 [7] questions the assumption that rigid tonal prescriptions applied to the *tz'u* before it became something of a lapidary craft in the twelfth century. Ultimately such prescriptions certainly were accepted.

As for the " large number of variant patterns, each of which bears the name of a musical air," *Tz'u-p'u* lists 826 *tiao-ming* 調名 or " tune titles " (*Tz'u-lü* gives 875) and illustrates them with specimens of 2,306 variant forms (*t'i*), ranging from 20 to 240 characters. Of some *tiao* only one form occurs, of others as many as a dozen or more.

Tiao of course means a tune, but the T'ang and Sung tunes associated with the *tz'u* have all disappeared.[8] The various *tz'u-p'u* contain no musical notations but present only the words, or what the music-publishing business calls " lyrics." Only the title of the tune is given (plus whatever information has been handed down about the musical mode to which it belonged, its derivation,

---

[6] According to Kondō Moku's 近藤杢 *Shina gakugei daijii* 支那學藝大辭彙 (Tōkyō, 1940) 445, the expression *t'ien-tz'u*, meaning to fill in characters after an existing model with a fixed pattern of tones, rhymes, and number of characters, was first used by the Ming writers Wu No 吳訥 and Hsü Shih-tseng 徐師曾. (This does not mean that the practice was not followed much earlier.) It became the standard term for the literary process of writing a *tz'u*.

[7] " Ssu-sheng shuo " 四聲說, *T'ung-sheng yüeh-k'an* 同聲月刊 1.6 (May, 1944) .1-8.

[8] *Tz'u* specialists have a generally accepted tradition for the singing of many of the old pieces, but it is not based on transmittal of T'ang or Sung airs. The only surviving *tz'u* accompanied by musical notations are in the *Po-shih tao-jen ko-ch'ü* 白石道人歌曲 (*SPTK* ed.) of Chiang K'uei 姜夔 (?1150-?1230), and attempts to decipher the system he used remain conjectural. Cf. Hsia Ch'eng-tao 夏承燾, " Po-shih ko-ch'ü p'ang-p'u pien " 白石歌曲旁譜辨, *Yen-ching hsüeh-pao* 燕京學報 12 (Dec., 1932) .2559-2588; his conclusions as to pitch are indicated in western musical notation by John H. Levis, *Foundations of Chinese Musical Art* (Peiping, 1936) 61. Ch'ien Wan-ch'ien 錢萬遷 has attempted a complete western-style score for one of Chiang K'uei's *tz'u* in " Ko-ch'i mei-ling ch'ü-p'u shuo-ming " 高溪梅令曲譜說明, *T'ung-sheng yüeh-k'an* 1.10 (Nov., 1942) .1-4.

alternate names, etc.), not the actual melody.[9] It is therefore more practical to translate *tiao* as "tune-pattern" or simply "pattern," since it is used to refer to the rhythmic, tonal, and rhyme schemes of the song-words, which can give us at most only a general notion of the outline of the melody.

The text of a given *tz'u* may or may not have some relation to the title of the musical air with which it is associated. It is as if we should call "Greensleaves" any poem modeled on the length, meter, and vowel distribution of the original words to that tune. In its early days as a pure song-form the *tz'u* was typically simple in content, and hardly more varied than the songs we hear on the radio today. In Northern Sung times, as poets using the *tz'u* primarily for literary purposes diversified their subject matter, they added subtitles. Su Shih's 蘇軾 "Nien-nu chiao—Ch'ih-pi huai-ku" 念奴嬌赤壁懷古 indicates both form and subject, as does "Sonnet: On the Extinction of the Venetian Republic." [10]

In partial explanation of some of the tune titles which are identical except for the addition or lack of an extra character such as *ling* 令, *yin* 引 or *chin* 近, and *man* 慢, it may be noted that these were musical terms which apparently indicated the length and tempo of the composition, though their precise relations to each other are not clear. The original melody might undergo numerous extensions, repetitions, and elaborations; in these successive forms it might take on totally new names, or

---

[9] LEVIS believes the tonal sequence of any Chinese poem intended for music indicates the progress of the melody upward, downward, or on a level. If this was ever true, it is not borne out by modern practice in singing, as a glance at a Chinese Christian hymnal will quickly show. At any rate, such a guide as LEVIS suggests might be sufficient for bardic chant as we conjecture it, or for the kind of *Sprechstimme* used by the late composer Arnold SCHÖNBERG, but not for singing in the usual sense. It is possible that T'ang and Sung performances of *tz'u* were more in the nature of *Sprechstimme* than of song as we generally think of it. The familiar Chinese "opera," indeed, is not characterized by *bel canto*.

[10] Cf. *Tz'u-p'u* 28.8ab. This is Su Shih's reworking, in *tz'u* form, of the first of his two famous *fu* on the Red Cliff. There are translations of this *tz'u* by CH'U Ta-kao (*Chinese Lyrics* [Cambridge, 1937] 24) and by WONG Man (*Poems from China* [Hong Kong, 1950] 85-86). Obviously we must go to the subtitle for the subject of the poem, for the first caption has nothing to do with it, being the name of a tune associated with a popular singing-girl of the time of Emperor Hsüan-tsung (713-755).

might retain the old name plus one of the above terms.[11] Investigation into the early phases of the *tz'u* requires careful consideration of a number of modern studies in Chinese and Japanese, various earlier *shih-hua* and *tz'u-hua*, and, of course, the texts of the early songs.[12]

The *tz'u* was evolved during the T'ang dynasty, and began to take its characteristic uneven shape in the ninth century. In T'ang times any words which were sung were *ko-tz'u* 歌 | or *ch'ü-tzu tz'u* 曲子 |—song-words—and these terms indicated the function of such words, rather than denoting a literary genre.[13] During most of the T'ang dynasty virtually all recorded song-words—

[11] These terms and others used in connection with *tz'u* forms and techniques are discussed, in somewhat statistical fashion, in the first chapter of FENG Shu-lan's *La technique et l'histoire du ts'eu* (Thèse pour le doctorat de l'Université de Paris, 1935).

[12] Although the material which follows is drawn from various sources, I am especially indebted to the authors of the following studies:

SUZUKI Torao 鈴木虎雄, "Shigen" 詞源, in his *Shina bungaku kenkyū* 支那文學研究 (Kyōto, 1922) 459-478.

AOKI Masaru 青木正兒, "Shikaku no chōtanku no hattatsu no genin ni tsuite" 詞格の長短句の發達の原因に就て, in his *Shina bungei ronsō* 支那文藝論藪 (Tōkyō, 1923) 67-85.

LIU Yün-hsiang 劉雲翔, "Wu-ko yü tz'u" 吳歌與詞, *T'ung-sheng yüeh-k'an* 2.2 (Feb., 1942).119-134. (This article is a reprint, without noticeable change except for the use of what apparently is the author's *hao*, of an article by LIU Yao-min 劉堯民 in *Kuo-li chung-yang ta-hsüeh pan-yüeh-k'an* 國立中央大學半月刊 [Nanking] 2.5[Dec., 1930].67-92].

HU Shih 胡適, "Tz'u ti ch'i-yüan" 詞的起原, reprinted as an appendix to his *Tz'u-hsüan* 詞選 (Shanghai, 1937) following page 381, and paginated independently.

[13] Some such designation of genuine song-words was needed, because by T'ang times the term *yüeh-fu* had been preëmpted by purely literary variations on song themes. just as the various cognates of the old Provençal *sonet*—any song—came to mean a lyric poem and eventually, in Italy and elsewhere, a specific form. Modern writers have sometimes used the word "sonnet" in a less restricted sense; MEREDITH called his "Modern Love" a sonnet sequence, although his stanzas departed from the fourteen-line form crystallized after Petrarch. Yet MEREDITH was not going back to archaic usage; he was not writing songs. Po Chü-i, when he wrote his "new *yüeh-fu*," returned to freer line-lengths than those in general use for literary *yüeh-fu* in his day; yet his introduction to these didactic pseudo-folk-poems indicates that he was not writing real songs to real tunes, and did not expect the poems to be sung. Although the term *yüeh-fu* has been used by many writers, especially in book titles, to cover virtually all kinds of real songs and pseudo-songs from those officially collected in Han times to those adapted in the early Yüan drama, it is useful for the purposes of this discussion to make a categorical distinction between T'ang *yüeh-fu*, which were not necessarily to be sung, and *ko-tz'u*, which were.

that is, lyrics actually intended to be sung, rather than literary mutations of obsolete song-forms—were in symmetrical stanzas with lines of equal length. The problem is to determine how *ch'ang-tuan chü* 長短句, long-and-short verse, which had become the most noticeable feature of song-words by the time *tz'u* was recognized as a distinctive musico-literary form, came about. The term *ch'ang-tuan chü* is one of several [14] often used as synonyms of *tz'u*. It will be in the interest of clarity to use the term *ch'ang-tuan chü* in the present study where the matter of unequal line lengths is emphasized.

As HIGHTOWER has pointed out,[15] " Poetry with unequal lines is as old as the *Classic of Songs*; examples occur in the *Ch'u tz'u* and among the Han dynasty *yüeh-fu*, where they are definitely associated with musical settings." During the Six Dynasties period the tunes of the Han songs disappeared, and poets who elaborated on their subject matter usually wrote in quite regular forms, progressively anticipating in practice the theories of tonal and formal harmony developed in the fifth and sixth centuries.

However, songs to new tunes (also referred to, loosely, as *yüeh-fu*) were often in free forms. From the Six Dynasties period may be cited examples both from the works of literary men and from anonymous popular song-words which have survived. Sometimes the irregularities are slight, as in the " Mei-hua lo " 梅花落 [16] of PAO Chao 鮑昭 (d. 466, Sung dynasty), a mixture of five- and seven-word lines (5-5-5-7-5-7-7-7). His " Yeh tso yin " 夜坐吟 [17] (7-7-3-3-3-3-3-3-3-3) is a bit more suggestive of the much later *tz'u* forms. SHEN Yo 沈約 (441-513, Ch'i and Liang dynasties) wrote a series of " Liu-i shih " 六憶詩 [18] with one short line (3-5-5-5-5-5).

---

[14] Perhaps the most elegant is *ch'in-ch'ü wai-p'ien* 琴趣外篇, " careless diversions on the lute."

[15] See note 2.   [16] *YFSC* 24.1a.   [17] *YFSC* 76.4a

[18] Cf. his collected works, *Shen Yin-hou chi* 沈隱侯集 2.59ab (in *Han Wei Liu-ch'ao pai-san-chia chi* 漢魏六朝百三家集, 1879 reprint, *ts'e* 74). These four *shih* (not classed as *yüeh-fu*) might be considered five-word cinquains introduced by three-character phrases something in the nature of subtitles. However, in reading the poems aloud one feels that the three-word lines have a definite rhythmic function.

Long-short verses from the Liang dynasty (502-556) are not rare. They include Hsü Mien's 徐勉 " Ying-k'o ch'ü " 迎客曲 and " Sung-k'o ch'ü " 送客 | [19] (both 3-3-7-3-3-7); T'ao Hung-ching's 陶弘景 " Han-yeh yüan " 寒夜怨 [20] (3-3-7-7-7-3-3-3-3-5-5); Wang Yün's 王筠 " Ch'u fei yin " 楚妃吟 [21] (3-3-3-3-3-5-3-5-3-5-7-5-3-5-5-5); two " Ch'ang hsiang-ssu " 長相思 by Chang Shuai 長率 and two by Hsü Ling 徐陵 [22] (three of these are 3-3-7-3-3-5-5-5-5, one of Chang Shuai's 3-3-7-7-7-5-5); seven " Chiang-nan nung " 江南弄 by Hsiao Yen 蕭衍 (Wu-ti 武帝) [23] (7-7-7-3-3-3-3); and " Ch'un-ch'ing ch'ü " 春情 | by Hsiao Kang 蕭綱 (Chien-wen ti 簡文帝) [24] (7-7-7-7-7-7-5-5).

From the Sui dynasty (589-618) a slightly irregular lyric, " Yeh-yin chao-mien ch'ü " 夜飲朝眠 | [25] is attributed to Yang Kuang 楊廣 (the second Emperor, Yang-ti 煬帝), though ascriptions of some other long-short compositions to him have been discredited.[26]

Some Chinese and Japanese writers have regarded such pre-T'ang products as prototypes of the *tz'u*'s long and short verses. Yang Shen 楊慎 (1488-1559) quotes and comments on most of the specimens mentioned above in his *Sheng-an tz'u-p'in* 升菴 | 品,[27] along with other Liang verses in regular meters in which he finds the germs of certain later *tz'u-tiao*.

[19] *YFSC* 77.3a.   [20] *YFSC* 76.5b.
[21] *YFSC* 29.11a.   [22] *YFSC* 69.5ab.   [23] *YFSC* 50.1a-3b.
[24] Cf. his collected works, *Liang Chien-wen-ti yü chih* 梁簡文帝御製 (*ts'e* 67 in collection mentioned in Note 18 above) 2.64b-65a.
[25] See Yang Shen 楊慎, *Sheng-an ho-chi* 升菴合集 158.2b-3a.
[26] Most significant of these are eight " Wang Chiang-nan " 望江南, of which not only the title but also the form is identical with a well-known *tz'u* pattern (also called " I Chiang-nan " 憶 | |, " Meng Chiang-nan " 夢 | |, and " Hsieh Ch'iu-niang " 謝秋娘—see note 75). Various writers have taken the attribution to Yang-ti as a basis for dating the *tz'u* form from the Sui period. But the ascription is made in a *hsiao-shuo* 小說 of questionable date and provenience, *Hai-shan chi* 海山集, considered spurious by the Ssu-k'u editors. Yang Shen (see note 25) noted several specimens of " Wang Chiang-nan " attributed to Yang-ti in various *ch'uan-ch'i*, but observed that they were not in typical Six Dynasties language. The tune itself apparently dates from the early ninth century.
[27] Yang Shen's *Tz'u-p'in* comprises *chüan* 151-62 (in *ts'e* 63-67) of his collected works mentioned in note 25; see *chüan* 158. Mao Ch'i-ling 毛奇齡 (1623-1716) adopted Yang's views on these poems as proto-*tz'u* in his own *Tz'u-hua* | 話 (cf. his

Such lyrics as these, all written in the Yangtze region, apparently were related to the living body of popular song in South China during the centuries of division—the *Wu-sheng ko* 吳聲歌, " songs in the Wu dialect "—with which they are classified in *Yüeh-fu shih-chi.* Doubtless the poets mentioned above were influenced by popular song words in long-short verses such as those of " Hua-shan chi " 華山畿 (3-5-5-3-5-5-5-5-5-5-3-5-5-5) ,[28] " Chiao-nü shih " 嬌女詩 (5-5-4-5-4-5) ,[29] and " Ch'ing-ch'i hsiao-ku ch'ü " 青溪小姑 | (5-3-4-4) .[30] Songs like these were apparently still being sung all along the lower Yangtze in the first century of the T'ang dynasty, and they were officially classified among the types of music for various uses at court.[31]

Despite qualifications and reservations, everyone seems to agree that there is some kinship between songs of this sort and the *tz'u* which evolved later. In the *tz'u-ch'ü* section of *Ssu-k'u ch'üan-shu tsung-mu t'i-yao*[32] it is observed that " examining the Wu dialect songs of successive periods, we find that some of the verses are long, some short. They are usually of a delicate and yielding tonality, already approaching the *hsiao-tz'u* 小 | " —that is, the short and relatively simple *tz'u* which characterized the genre with T'ang and Five Dynasties poets.

Yang Shen and Mao Ch'i-ling seem to have assumed that T'ang popular and literary poets simply added to the variety of line-lengths they found in the Wu songs until they produced

collected works, *Hsi-ho ho-chi* 西河合集 1[*ts'e* 88].8ab). Liu Yün-hsiang (*op. cit.* 120) agrees, saying that the form of these Six Dynasties verses is essentially that of the later *tz'u,* and that their casualness and facility approach the " tone " of the *tz'u.* Aoki (*op. cit.* 83), however, maintains that it is just these qualities that T'ang *tz'u* lack, the latter showing a discipline of expression and pattern (resulting from the practice of regulated forms) that make their poetic " tone " quite distinct from that of these " Six Dynasties *yüeh-fu.*"

[28] *YFSC* 46.2b-4a gives twenty-five stanzas, from thirteen to twenty-three characters long. Thirteen of them are 5-5-5, ten are 3-5-5, one 5-5-5-5, and one 3-5-5-5.

[29] *YFSC* 47.6a.

[30] *YFSC* 5b-6a. This is the " Children's Song " familiar in Arthur Waley's translation (*170 Chinese Poems* 123).

[31] See note 37.

[32] Commercial Press edition (Shanghai, 1932) 4.4462 (entry on *Li-tai shih-yü* 歷代詩餘).

patterns which ultimately became known as *tz'u-tiao*. But whatever the significance of the Wu songs as a precedent, it appears
that long-short verse had virtually ceased to be written by the
beginning of the eighth century—or at least, that if it was being
produced in any quantity prior to Lı Po's non-musical *yüeh-fu*
experiments, it was regarded as subliterature and was not being
preserved.

There seem to have been two reasons for this, one musical and
the other literary. The increased importation of foreign music
under the cosmopolitan Sui-T'ang dynasties submerged or greatly
altered the songs of South China. On the literary side, the prestige of regulated verse (*lü-shih* 律詩), as finally codified in the
early eighth century, overshadowed that of all other types of
poetry.

The importation of music from Central Asia, Korea, and other
non-Chinese areas had been going on ever since the Han dynasty,
but particularly since the alien incursions of the fourth century.
Soldiers picked up foreign melodies while garrisoning the fron-
tiers or campaigning beyond them. Foreigners occupied Chinese
territories whose native inhabitants took over the music of their
conquerors. The Southern dynasties carried on intermittent com-
merce with the conquered Northern and Western regions, and
the traders also brought home new types of music. Buddhist
missionaries and pilgrims brought religious and secular melodies
from Central Asia and India.[33]

It would be rash to assume that the Chinese promptly dropped
their " own " music [34] entirely and adopted intact every foreign

---

[33] A notable combination of the military and religious factors was the marauding
expedition against Kucha (a highly cultured Aryan kingdom in the Tarim basin)
around 382 by Lü Kuang 呂光 (afterward founder of the " Later " Liang 涼
Kingdom), who brought back to China both the famous Buddhist translator Kumā-
rajīva and a Kuchan orchestra. Of course, music also moved from China outward.
For example, " Ch'in-wang p'o-chen yüeh " 秦王破陣樂 (cf. under its later name
" Ch'i-te wu " 七德舞 in *Shina gakugei daijii* 485), a song of victory by Lı Shih-min
李世民, was carried by his armies to the limits of his far-flung empire after he
mounted the throne as the second emperor of the T'ang. Under the Sui and T'ang,
envoys and religious pilgrims from Japan took back to their country music in both
older and newer styles. The sole survival of the elements of T'ang music today is
believed to be found in some of the court music of Japan.

[34] That is, what was considered to be their own music at the time.

tune they heard. Undoubtedly they modified the strange melodies to some extent, to suit their habitual ways of singing and playing. On the other hand, the exoticism of the new music formed perhaps its strongest appeal. The Chinese certainly made some changes in their traditional scales and musical devices when they adopted foreign instruments like the *p'i-pa* 琵琶, and altered many of their older tunes accordingly.[35]

By the time the empire was reunited under the Sui dynasty, foreign types of music were dominant among virtually all classes of society, and it is said there were few who still appreciated the older " native " styles.[36] Of the ten classes of music officially designated by T'ang Emperor T'ai-tsung all but two, *yen-yüeh* 燕樂 and *ch'ing-shang yüeh* 清商 |, were of foreign origin.[37]

The " songs in the Wu dialect " were part of the *Ch'ing-shang yüeh* class.[38] According to the *T'ung-tien* 通典 this whole class of

---

[35] A performer on the *p'i-pa*, which had only four strings, could not play all the tunes devised for the older seven-stringed *ch'in* 琴, or at least could not play them the same way.

[36] This long continued to be a standard subject for complaint even by writers who amply demonstrated their own fondness for currently popular tunes, such as Po Chü-i (cf. WALEY, *op. cit.* 185: " Ancient melodies—weak and savourless, Not appealing to present men's taste.")

[37] The others were Hsi-liang *yüeh* 西涼樂, T'ien-chu 天竺 *yüeh*, Kao-li 高麗 *yüeh*, Ch'iu-tz'u 龜茲 *yüeh*, An-kuo 安國 *yüeh*, Su-le 疏勒 *yüeh*, Kao-ch'ang 高昌 *yüeh*, and K'ang-kuo 康國 *yüeh* (SUZUKI, *op. cit.* 467), all bearing the names of non-Chinese territories stretching from Korea across the Tarim basin to India. (*Yen-yüeh* does not refer to the territory of the old state of Yen, but means "feast music," music for entertainment, in contrast to the older formal, ceremonial music— *ya-yüeh* 雅樂—of state functions.) Music of foreign origin was in use at the courts of the contemporary (Southern) Ch'en 陳 and (Northern) Chou 周 dynasties, received official status under their successor the Sui (see L. C. GOODRICH, " Foreign Music at the Court of Sui Wen-ti," *JAOS* 69.3[1949].148-49), and was firmly established under the first T'ang emperors. Kao-tsung specified nine types (*pu* 部) of music based on Sui regulations; T'ai-tsung eliminated one and added two, one of which he called *yen-yüeh pu*. SUZUKI however thinks that all ten types could be called *yen-yüeh* (= 宴 |) in the general sense.

[38] Cf. *Shina gakugei daijii* 688, entry on " Ch'ing-shang-ch'ü ko-tz'u " 清商曲 歌辭. Cf. also the section so entitled in *YFSC*, comprising *chüan* 44-51, of which the " Wu-sheng ch'ü-tz'u " 吳聲曲辭 occupy *chüan* 44-46 and part of 47. The other subdivisions are " Hsi-ch'ü-ko " 西曲歌, " Yüeh-chieh che yang-liu ko " 月節折楊柳歌, " Chiang-nan nung " 江南弄, " Liang ya-ko " 梁雅歌, and " Shang-yün yüeh " 上雲樂.

songs, music and words alike, had fallen into utter desuetude by
the middle of the eighth century. In the section on music [39]
it is stated that from the Ch'ang-an 長安 period (701, at the
close of the reign of the Empress Wu) the old tunes were no
longer esteemed at court, and musicians neglected them. Only
eight tunes are listed as being playable at that time, and it is
said that so many of the words had been corrupted or forgotten
that the songs now bore scant resemblance to the old Wu *melos*.
Someone suggested that Southerners should be encouraged to
practice and transmit their old songs, but apparently little came
of the proposal. In the K'ai-yüan 開元 period (713-741) a
Northerner called Lı Lang-tzu 李郎子 claimed to have studied
the old songs with a Southern master. According to *T'ung-tien*,
after the death of Lı both the instrumental and vocal parts of the
Ch'ing-yüeh (or *Ch'ing-shang yüeh*) were wholly neglected.

Tu Yu 杜佑, who compiled the *T'ung-tien*, died less than eighty
years after the K'ai-yüan period, and there is no reason to dis-
credit his information. It may be that it reflects primarily the
musical scene at the capital—which was now in the Northwest
at Ch'ang-an, far removed from the seats of the Six Dynasties—
and that while the Wu songs may have passed out of fashion at
court and among the *haut monde*, provincial Southerners may
have continued to sing them. It is at least conceivable that
popular songs with some irregular line-lengths continued (though
no doubt with changes of style and convention) in an unbroken
stream in the South until the *tz'u* itself emerged as *ch'ang-tuan
chü* in the ninth century.[40]

---

[39] *T'ung tien* (Che-chiang Shu-chü 浙江書局 edition of 1896) 146 (*han* 4, *ts'e*
7) .2b-3a.

[40] Of the numerous popular ballads and children's songs recorded in the "Wu-hsing
chih" 五行志 of *Hsin T'ang shu*, most are in even quatrains or couplets; but a few
are quite assymetrical, such as the following (25.10b) said to have been current in
the T'ien-pao 天寶 period (742-755):

燕燕
飛上天
天上女兒鋪白氈
氈上有千錢

It is certain, however, that the old Wu songs had ceased to be sung at Court and had ceased to be imitated by literary poets. We may assume then that the long-short verse which often characterized them did not serve as model for the lyrics sung to the foreign-style tunes in the first two centuries of the T'ang. These tunes multiplied both by further importation and by adaptation or imitation in the Chinese musical profession, above all, in the imperial music-factory, the *chiao-fang* 教坊.[41]

Although many real foreign tunes were current, the foreign words to them, of course, were not. Many tunes must have become popular in purely instrumental form. But just as any popular instrumental piece of music, if singable, is likely to have words written to it—either more or less spontaneously among the people, or by professional purveyors of entertainment (*Finlandia* becomes " Dear Land of Home," RAVEL's *Pavane* " The Lamp is Low ")—these tunes acquired Chinese words. Often

---

Fly swallow fly
Up to the sky—
The Girl in the Sky has spread her white rug,
On it are a thousand cash.

Such living songs, as well as the remote precedent of the Han *yüeh-fu*, may have influenced LI Po and Po Chü-i in the versification of their neo-old-style folk poems. LI Po for instance made telling use of an occasional three-word line. These literary experiments, however, were not written to any known tunes, and there is no evidence that any of them were ever set to music.

[41] This office, something like the Han *yüeh-fu*, was established by Hsüan-tsung in 714 for the collection and preservation of songs and dance music both formal and popular, as well as for the composition and performance of new music for various occasions. It is not to be confused with his private " theater," the famous Pear Garden (*Li-yüan* 梨園), though the latter presumably made use of the *chiao-fang* material and possibly some of its personnel. With various modifications of function the office lasted under the name of *chiao-fang* down to early Ch'ing. For its organization under the T'ang see the " Po-kuan chih " 百官志 of the Old and New *T'ang shu*; for interesting musical and anecdotal material see *Chiao-fang chi* 教坊記 (*TTTS, ts'e* 5). The author of the latter, Ts'ui Ling-ch'in 崔令欽, apparently lived in Hsüan-tsung's reign (see HU Shih, *op. cit.* 19), but a list of some three hundred tune titles appended to the book contains names of tunes elsewhere reported to have originated considerably later, such as " Wang Chiang-nan " (see note 75) and " P'u-sa man " (see below, p. 00). Either the list was not part of the original text or it contains later interpolations. Dr. HU therefore, in an exchange of letters with WANG Kuo-wei 王國維 (see *Tz'u-hsüan* 18-21), maintains that this list cannot be used to prove that such-and-such a tune existed in the first half of the eighth century.

they acquired not one set of words but several (as RACHMANI-
NOV's C-minor piano concerto becomes in one version " Full Moon
and Empty Arms," in another " And Still the Volga Flows," and
as *Londonderry Air* becomes both " Danny Boy " and " Would
God I Were the Tender Apple Blossom ") .

Since the metrics and musical phrasing of these new tunes
were different from those of traditional Chinese music, one might
expect that the words now written to them would not be bound
by standard Chinese practice in the composition of literary verse
at the time, which was concentrated on strictly regulated forms.
The fact is, nevertheless, that the surviving Chinese texts associated with these tunes in the first two centuries of the T'ang era
are all symmetrical, and nearly all " regulated," forms. They are
invariably four-line stanzas—of six, oftener five, and most often
seven-word lines, sometimes written for a specific tune but in
many cases borrowed from the works of well-known poets.[42]
Occasionally part of a *ku-shih* 古詩 was used, but more commonly
half of an eight-line *lü-shih* was adapted. Most popular of all was
the *chüeh-chü* 絕句, and the portions of other poems selected
were so similar to the self-sufficient *chüeh-chü* that it will be
practical here to refer to all such quatrains as *chüeh-chü*.

For a short, simple tune a single *chüeh-chü* would suffice. For
longer compositions (*ta-ch'ü* 大曲) which fell into sections with
contrasting tempi or rhythms, sequences of these four-line stanzas
were sung, separated by instrumental interludes.[43]

[42] SUZUKI (*op. cit.* 467-69) identifies a number of these. The words to " Kai-lo
feng " 蓋羅逢 were those of WANG Ch'ang-ling's seven-word *chüeh-chü* " Kuei yüan "
閨怨 beginning 秦時明月漢時關. The words to " K'un-lun tzu " 崑崙子 were
the first half of WANG Wei's five-word *lü-shih* entitled " Ts'ung Ch'i-wang kuo Yang-
shih pieh-yeh " 從岐王過楊氏別業, and the words to " Jung hun " 戎渾, the
last half of his five-word *lü-shih* " Kuan lieh " 觀獵. The first stanza of " Lu-chou ko "
陸州歌 was the last half of WANG Wei's famous five-word *lü-shih* " Chung-nan
shan " 終南山, with the change of one word. The third stanza of this *ta-ch'ü*
utilized the first four lines of a *ku-shih* by KAO Shih 高適 beginning 哭單父梁
少府, again with a few words changed. (Where KAO Shih wrote 子雲居 the words
of the *ta-ch'ü* are recorded as 紫雲車, obviously a transformation wrought through
the ear rather than the eye.)

[43] The words to these *ta-ch'ü* may be examined in *YFSC* 79, or in *Ch'in-ting
tz'u-p'u* 40.

Tz'u-p'u relegated the *ta-ch'ü* to its last *chüan*, presumably because they were not

Even extemporaneous song-words, improvised on the spot for some occasion, were composed in regular quatrains. *Pen-shih shih* 本事詩 [44] gives an anecdote of the court of Chung-tsung 中宗 around the year 708, when at a feast the Emperor's guests in turn sang verses to the tune "Hui-po yüeh" 回波樂. Those of Li Ching-po 李景伯 and Shen Ch'üan-ch'i 沈佺期 are given, and both are in four six-word lines.[45]

Also in four six-word lines are the six surviving stanzas of "Wu-ma tz'u," 舞馬 |, which Chang Yüeh 張說 wrote for a sort of ballet of horses performed at a celebration of the Emperor Hsüan-tsung's birthday.[46] This song-sequence serves the function

---

used as models by later *tz'u* writers. However, many of their titles, slightly altered (e. g., "Liang-chou ling" | | 令 instead of "Liang-chou ko"), were used for later *ch'ang-tuan-chü*. The foreign origin of the tunes is often hinted in their titles; many bear the names of the northwestern districts of medieval China, through which these non-Chinese melodies streamed in.

In the *ta-ch'ü* the quatrains are grouped in sections corresponding to the musical divisions of the composition; the first group is the *ko* 歌, followed by a section called *ju-p'o* 入破 or *p'ai-pien* 排遍, according to its length and tempo (Suzuki, *op. cit.* 468). Both seven-word and five-word quatrains often appear in a single *ta-ch'ü*. According to Suzuki an eleven-stanza pattern was more or less standard for a full-length *ta-ch'ü*: the *ko* in five stanzas and *ju-p'o* in six, of which the last was called the *ch'e* 徹.

[44] Meng Ch'i 孟啓 (T'ang) here gathers anecdotes purporting to explain the circumstances of the writing of various poems. The stories are ranged under seven categories of which the last, Satire (嘲戲), includes the "Hui-po yüeh" incident (*TTTS, ts'e* 6, 8a).

[45] Suzuki (*op. cit.* 464-65) suggests that the tune and subject matter of "Hui-po yüeh" (about the pleasures of song and dance) are related to the sixth-century "Kao-chü-li ch'ü" 高句麗曲 and "Huan t'ai yüeh" 還臺樂, of which surviving lyrics by Wang Pao 王襃 (Northern Chou) and Lu Ch'iung 陸瓊 (Ch'en) respectively are not in four but six lines of six words each. The guests of Chung-tsung's court still followed the tradition of six-word lines, but at the same time, in using only four such lines, they also followed the growing trend toward the *chüeh-chü*. It is worth noting that Shen Ch'üan-ch'i was one of the poets credited with explicit codification of the practices of the *lü-shih*.

[46] *Chang Yüeh-chih wen-chi* 張說之文集 (*SPTK* ed.) 2.9ab. Two of the stanzas are given in *Tz'u-p'u* 1.12ab, where an introductory note describes the spectacle as reported in the treatises on ceremony and on music in the T'ang histories. The performance is said to have been a yearly affair. The *corps de ballet* is reported to have consisted of no less than "four hundred hooves." The horses were caparisoned in rich embroideries, with gilt and jeweled halters; they lifted their hooves, tossed

199

of the *tz'u* as well as using the character in its title, since patently
it was written for use with specific music, to which the horses
were trained. One is not surprised that the words follow an even
rhythmic pattern, for the music must have done so too, even
though the horses doubtless were more agile and adaptable than
the elephants which had so much trouble with STRAVINSKY's music
for the Ringling Brothers. In any event the regularity of the
literary form was in accord with poetic fashion. Hsüan-tsung's
reign was in the heyday of the *chüeh-chü*, when major poets like
WANG Ch'ang-ling 王昌齡, WANG Wei 王維,[47] and LI Po 李
白 [48] were turning out by the score quatrains esteemed not by
posterity alone, but as popular in their own day as the lyrics of
Cole PORTER and Oscar HAMMERSTEIN in ours.

their heads, and switched their tails in time to the music, which went on through
several dozen choruses. Some of them danced on platforms supported by muscle-men.
(These feats are only mildly impressive compared to those of John BANKS' horse
which in the late sixteenth century danced on the roof of St. Paul's. BANKS' horse
could also add and subtract, and could "tell maids from mawkins," for which
accomplishments BANKS rashly took it to Rome where the Pope condemned it to be
burned as a witch.)

[47] WANG Wei's seven-word *chüeh-chü* "Sung Yüan-erh shih An-hsi" 送元二使
安西, addressed to a friend who was going on a mission to Central Asia, became so
popular as a parting song that it is classed as a *yüeh-fu* in many collections under the
title "Wei-ch'eng ch'ü" 渭城曲, and *Tz'u-p'u* (1.28a-29a) includes it as a *tz'u-tiao*,
"Yang-kuan ch'ü" 陽關曲. Under this and similar titles (e. g., "Yang-kuan san-
tieh" | | 三疊), and with various additions and repetitions, it has been used for
over a thousand years as a song of farewell. Later poets metamorphosed it into a
ch'ang-tuan-chü (see "Yang-kuan yin" | | 引, *Tz'u-p'u* 18.23b). The tonal pattern
of WANG Wei's poem and of later ones strictly modeled on it (including two by SU
Shih) is discussed in *Tz'u-p'u*, and at greater length by MORI Taijirō 森泰次郎—cf.
Tso-shih-fa chiang-hua 作詩法講話 (translations of some of MORI's lectures by
CHANG Ming-tz'u 張銘慈, Shanghai: Commercial Press, 1931) 37-40. According to
Tz'u-p'u, CH'IN Kuan 秦觀 (1049-1100) wrote that in his time "Wei-ch'eng ch'ü"
was sung to the tune of "Hsiao Ch'in wang" (see p. 126), which *Chiao-fang chi*
identifies as the same as "Ch'in wang po-chen yüeh" (see note 33).

[48] A tradition originating in Sung times (see SUZUKI, *op. cit.* 475, quoting *Hsiang-
shan yeh-lu* 湘山野錄 by the Sung Buddhist writer Wen-ying 文瑩; and LIU
Yün-hsiang, *op. cit.* 123, citing *T'ung-chih* 通志) and questioned in Ming (SUZUKI
quotes HU Ying-lin 胡應麟, *Chuang-yüeh wei-t'an* 莊嶽委譚) made LI Po the
originator of the long-and-short *tz'u*. *Ts'un-ch'ien chi* 尊前集, one of the early *tz'u*
anthologies, formerly thought to have been put together about the same time as
Hua-chien chi, but probably of Sung date (see SHIZUKU Sekkō 雫石晧, "*Sonzenshu*

The T'ang poets heard their words being sung not only at court but in the wine shops and brothels, in the provinces as well as in gay Ch'ang-an. An anecdote in *Chi-i chi*[49] 集異記 illustrates this prompt mating of poetry and song. WANG Ch'ang-ling, KAO Shih 高適, and WANG Chih-huan 王之渙 were having supper in a pleasure-house when a group of actors from the Emperor's theater arrived for a banquet, followed by sing-song girls. When the music started the three poets agreed among themselves to determine their relative merits by the number of their poems they would hear sung at the party. By and by an actor sang a *chüeh-chü* of WANG Ch'ang-ling's, and another sang one by KAO Shih. " These vulgar actors," said WANG Chih-huan, " what do they know? It's their nature to prefer the familiar to the exquisite. But I'm sure that when it comes the turn of the singing-girls, the most beautiful will sing one of my songs." And so she did.

zakkō" 尊前集雜考, *Kangakkai zasshi* 漢學會雜誌 9 [Tōkyō, June, 1941].97-106) attributed twelve *tz'u* to LI Po, and *Ch'üan T'ang shih* gives him fourteen. Both collections include as *tz'u* the three " Ch'ing-p'ing tiao " 清平調 about YANG Kuei-fei 楊貴妃 which, according to tradition, offended the lady and led to the poet's departure from court. These are said to have been improvised to a tune combining two modes, and straightway sung by the famous vocalist LI Kuei-nien 李龜年 while Hsüan-tsung himself played the tune on a jade flute. Of the " *tz'u* " mentioned above these are the only ones which appear in the earliest editions of LI Po's poems after his death, and they are three quite regular *chüeh-chü*. We need not doubt that they were written as *ko-tz'u*, and *Tz'u-p'u* (40.1a) places them at the beginning of its *ta-ch'ü* section. They certainly are not *ch'ang-tuan-chü*, as are the other titles attributed to LI Po in the anthologies mentioned. These are in five patterns, " P'u-sa man " 菩薩蠻, " I Ch'in o " 憶秦娥, " Ch'ing-p'ing yüeh " 清平樂 (not to be confused with " Ch'ing-p'ing tiao "), " Lien-li chih," 連理枝, and " Kuei-tien ch'iu " 桂殿秋. A specimen of each of these, except the last, appears in *Tz'u-p'u* under the name of LI Po. However, LIU (*op. cit.* 123), SUZUKI (*op. cit.* 475), and HU (*op. cit.* 2) adduce various evidence that these attributions are false. LI Po certainly could not have written a poem for the tune " P'u-sa man," which originated around the middle of the ninth century, long after his death (see p. 00). As a matter of fact one of the " P'u-sa man " attributed to LI Po in *Ts'un-ch'ien chi* is the well-known song in praise of Chiang-nan by WEI Chuang 韋莊, who lived into the tenth century.

LI Po did indeed write what he called *yüeh-fu* in verses often of variable and irregular lengths, but these were literary experiments, not songs to music.

[49] This account of brief and often incredible incidents, by HSIEH Yung-jo 薛用弱 (fl. 830), appears in several collections including *T'ang-tai ts'ung-shu*, *ts'e* 10; for the story referred to, cf. folio 34a. It is retold by John C. H. WU in " The Four Seasons of T'ang Poetry," *T'ien Hsia Monthly* 7.4 (Shanghai, 1938).358-59.

Chi-i chi in recounting the story refers to these song-words as *tz'u* and *ko-tz'u.* The poems are all quite regular *chüeh-chü,* as are virtually all extant song-words of that time. As remarked earlier, no distinction as to form was then thought of between *shih* and *tz'u,* the latter term designating not a literary genre but simply the function of the poem as song-words. Even many *ku-shih* could be sung, and apparently all *chüeh-chü.*

Now since all T'ang regulated verse maintained a constant rhythm in lines of equal length, a reader who had never seen any comment on the music of the time would suppose that the tunes to which such verse was sung would be likewise four-square in its periods, like the symmetrical music of eighteenth-century Europe. But unless Chinese musicians had transformed the Central Asian and other foreign elements beyond recognition, and had as well completely renounced the tradition of rhapsodical irregularity in earlier Chinese music, such as that of the Songs of Wu, such cannot have been the case.

It might be conjectured then that *chüeh-chü* were set only to the portion or portions of a musical composition which did happen to be melodically symmetrical, in other words where the poem happened to fit the music, leaving perhaps the introduction, one or more interludes, and finale to the instruments alone. In the case of the *ta-ch'ü* we know there were such purely instrumental sections. But this does not fully explain the manner in which poems were applied to music, for there is ample evidence that the words were adapted and supplemented in various ways.

Even in the most symmetrical vocal music of the West we do not expect an unvarying word-to-note, or even syllable-to-note, correspondence of text to music. A single syllable may take the shape of a melodic turn or phrase (by what in musical terminology is called melisma), or may be held by the voice on a single note while the instruments execute melodic or harmonic progressions. Or the voice may be given a short or long " rest " while the instruments play on. Often a line of verse is made to conform to the length of the melodic line by repetition of a word or phrase once or several times, or by the use of such interjections as " oh," " ah," " ohimè," " hélas." And how often, when the singer has had no actual words to sing, has he filled in with " tra-la-la," " hey-nonny-nonny," or more recently " vo-do-de-o-do."

There is reason to believe that T'ang singers made use of most or all of these devices, as the performers in the Chinese " opera " have certainly done down to the present day.[50] According to the most commonly held theory, the expedient of interpolated words or sounds was the major factor in the evolution of T'ang song-words from *chüeh-chü* to *ch'ang-tuan chü*.

Chinese writers from Sung to Ch'ing have used various terms in referring to these interpolations: *ho-sheng* 和聲 , *hsü-sheng* 虛 ｜, *fan-sheng* 泛 ｜, *san-sheng* 散 ｜. It is debatable whether any clear-cut distinctions should be or can be made between these terms as to the type or function of the interpolations meant— e. g., emotional interjections, meaningless vocalizations, refrains by the singer, chorus with hand-clapping by the orchestra or the audience. Further on in this study they will be referred to indiscriminately as " expletives "; in the following quotations, the Chinese terms are simply transliterated.

Writing on the relation of poetry to music in earlier times, the versatile astronomer and mathematician SHEN K'ua 沈括 (1030-1094)[51] stated:

. . . Aside from the verse-text (詩) there were also *ho-sheng*. What we call songs (曲), in the case of the old *yüeh-fu* always had both notes (聲) and text (詞). When these were written together, " ho-ho-ho " (何 ｜｜ or 賀 ｜｜) and the like were all *ho-sheng*. The *ch'an-sheng* 纏 ｜ (grace notes? connecting passages?) in the music of today are devices which stem from these. In T'ang times, people began to write words directly to music. This form is said to have begun with WANG Yai 王涯 (?764-825),[52] but many followed it in the Chen-yüan through Yüan-ho 元和 periods (785-806).

HU Tzu 胡仔 (*ca.* 1147) wrote:

Early T'ang song-words were mostly five-word *shih* or seven-word *shih*;

---

[50] The *nan-ch'ü* 南曲 or " Southern drama " term for melisma is *mo-tiao* 磨調 (*Shina gakugei daijii* 1220). AOKI (*op. cit.* 76) and HU (*op. cit.* 12) speak of the frequency of " helping words " or " ornamental words " (*ch'en-tzu* 襯字) in the Yüan and later drama. See note 60 below.

[51] Cf. his *Meng-ch'i pi-t'an* 夢溪筆談 (*SPTK* ed.) 5.9ab.

[52] All WANG Yai's *shih* and *yüeh-fu* in *Ch'üan T'ang shih* (*han* 2, *ts'e* 13) 1a-2a are in quite regular forms.

there were no long-and-short verses. From the Middle [T'ang] period [53] on into the Five Dynasties, they evolved into *ch'ang-tuan chü*, until in our own time this form is practiced generally. [Of the older *tz'u* complete with words and music?] still extant, the two pieces " Jui che-ku " 瑞鷓鴣 and " Hsiao Ch'in wang " 小秦王 are simply an eight-line *shih* and a seven-word *chüeh-chü-shih* [respectively]. " Jui che-ku " still may be easily sung in accordance with the words. In the case of " Hsiao Ch'in wang " it is necessary to mix in *hsü-sheng* in order to sing it.[54]

CHU HSI 朱熹 (1130-1200) explicitly stated the theory that the use of such extra-textual sounds or words crystallized into irregular verse forms. " The old *yüeh-fu* were simply *shih*," he is reported as saying, " into which a number of *fan-sheng* were inserted. Later, people hesitated to omit these expletives, so for each one they inserted an actual character, producing long-and-short verse. Thus the songs (曲) of today." [55]

The prestige of any observation of CHU HSI's may have influenced later writers on the subject; at any rate, the theory has the appeal of any neat categorical explanation. And it fits well with a term used since Southern Sung times as a synonym for the *tz'u*: *shih-yü* 詩餘 implies that the form was an extension or outgrowth of the *shih*.[56]

Ch'üan T'ang shih 全唐詩 (1703), adapting CHU HSI's explanation in a note at the beginning of the appended *tz'u* section,[57]

---

[53] T'ang poetry is generally divided into four periods as standardized after KAO Tai's 高棣 *T'ang shih p'in-hui* 唐詩品彙 (Ming): Early (初) T'ang, from 618 to the accession of Hsüan-tsung in 713 (a total of 115 years); Developed or Full (盛) T'ang, 713 to 766, shortly after the death of LI Po (53 years); Middle (中) T'ang, 766 to 836, a dozen years after the death of HAN Yü and a decade before that of Po Chü-i (70 years); and Late (晚) T'ang, 836 to the fall of the dynasty in 907 (71 years). These western-style dates encompass successive reign-periods. See *Shina gakugei daijii* 418.

[54] *T'iao-ch'i yü-yin ts'ung-hua* 苕溪漁隱叢話 (in *Tz'u-hua ts'ung-pien*), sec. 39 長短句, 5b.

[55] *Chu tzu yü-lei* 朱子語類 (1876 ed.) 140.9a.

[56] As in the titles of the anthology *Ts'ao-t'ang shih-yü* 草堂詩餘 (late twelfth century) and of HUANG Chi's 黃機 collected *tz'u*, *Chu-chai shih-yü* 竹齋詩餘. See the Table of Contents of CHU Tsu-mou's 朱祖謀 *Ch'iang-ts'un ts'ung-shu* 彊村叢書 (Shanghai, 1922) for numerous later *tz'u* collections entitled—*shih-yü*.

[57] *Ch'üan T'ang shih*, han 4, ts'e 8, 54a. Rather than referring back to SHEN K'ua, possibly a clearer indication of precisely what the editors meant by *ho-sheng* is given in a book finished twenty-six years before *Ch'üan T'ang shih*, WAN Shu's 萬樹

reverts to the term *ho-sheng*: " T'ang *yüeh-fu* at first used such *shih* [forms] as *lü*[-*shih*] and *chüeh*[-*chü*], which were sung by mixing in *ho-sheng*. When for these expletives actual characters were written, lengthening and shortening the lines conforming to the rhythm of the tune, this was ' filling in (*t'ien-*) *tz'u*.' "

Other Ch'ing writers have said more or less the same thing. FANG Ch'eng-p'ei 方成培 (early nineteenth century) differs only in that he uses a different term for the expletives:

In T'ang times people sang for the most part five- and seven-word *chüeh-chü*, which could be set to music only by mixing in *san-sheng*. This was a spontaneous expedient. In the course of time these expletives were recorded (譜), being realized with one or more characters and thereby bringing about long-and-short verse. . . . Thus it was the *tz'u* which relieved the exhaustion of the " modern style "[58] and carried on the evolution of the *yüeh-fu*.[59]

As indicated earlier, no very determined attempt will be made here to establish hard and fast distinctions between the -*sheng* compounds which appear in the foregoing quotations, but tentative distinctions may be suggested. *Hsü-sheng* would seem to indicate sounds without meaning, or at least without relevance to the sense of the song.[60] AOKI (*op. cit.* 68) equates CHU Hsi's

---

Tz'u-lü 詞律 (Preface, 1687). In 1 (*ts'e* 2) .1a-2a, are reproduced several " Chu-chih " 竹枝 *tz'u*. In all of them, each seven-word line is divided ⅘, with the phrase *chu-chih* in smaller characters after the fourth word, and similarly *nü-erh* 女兒 after the seventh. WAN Shu comments: " The *chu-chih* and *nü-erh* used were sounds with which, during the singing, the crowd joined in (乃歌時羣相隨和之聲), as with the *chü-cho* 舉棹 and *nien-shao* 年少 in ' Ts'ai-lien ch'ü ' 採蓮曲."

For " Ts'ai-lien chü " or " Ts'ai-lien tzu " see p. 129.

[58] *Chin-t'i* 今體 = *lü-shih* and *chüeh-chü*.

[59] *Hsiang-yen-chü tz'u-chu* 香研居詞麈 (in *Hsiao-yüan ts'ung-shu* 嘯園叢書, *ts'e* 31) 1.1ab.

[60] WU Heng-chao 吳衡照 (1771-?) in his *Lien-tzu-chü tz'u-hua* 蓮子居詞話 (*Tz'u-hua ts'ung-pien*) 1.8b-9a, comments on HU Tzu's term *hsü-sheng*: " These are what *Yüeh-fu chih-mi* 樂府指迷 speaks of as the musicians' and singers' *ch'en-tzu* 襯字. . . . They were a convenience to the singer, as in the old *yüeh-fu* [the words] ' fei-hu-hsi ' 妃呼豨 . . . . As a general rule seven-word *chüeh-chü* were always handled this way." The line " fei-hu-hsi " occurs in the anonymous (Han?) " Yu so ssu " 有所思 (cf. *Ku-shih yüan* 古詩源 [SPPY ed.] 3.13a) in which context it is unintelligible; WALEY's translation (*170 Chinese Poems* 55) omits it.

Ch'en-tzu (helping words or ornamental words) is a term associated with the later dramatic songs (particularly the *pei-ch'ü* 北曲) rather than with the *tz'u*, and

fan-sheng with them. Both he and SUZUKI (*op. cit.* 477) think *ho-sheng* as used in *Ch'üan T'ang shih* refers specifically to refrains (using real words, not vocalizations like "*i-a-na*" 伊阿 那) in which the instrumentalists, or possibly the audience, joined. *San-sheng* would seem to cover any kind of interpolation.

Why should T'ang singers have stuck so long to *chüeh-chü* for their basic song-words, since so much ingenuity was required to make them singable? Or conversely, why should musicians have stuck to tunes that did not fit the words? The answer seems to be that both were what the public wanted. On the one hand, a tune as four-square as a *chüeh-chü* would be too monotonous to bear much repetition. On the other, the literary prestige of *lü-shih* and *chüeh-chü*, as mentioned before, was supreme. It was great even before the rules of regulated verse were codified early in the eighth century, and when the court stipulated regulated forms for the state examinations, they became the stock-in-trade of every writer. The regularity of the *chüeh-chü*, its compactness, and its capacity for saying much in few words made it catchy and easy to remember. Probably nobody objected to the interpolated asides, refrains, or patter words; people have always liked such things in popular songs, and still do. However, customs and fashions in art and entertainment change, and eventually the *chüeh-chü*, adapted as a song-form, began to be transformed into something else.

It is assumed that the interpolations were written down as an aid to memory by the singers who originated them or borrowed

refers to short asides or "ad libs" outside the tonal pattern; see *Ch'in-ting ch'ü-p'u* 欽定曲譜 (1941 photo-reprint of Palace edition), Introduction, 6a.

SHEN I-fu 沈義父, the thirteenth-century author of *Yüeh-fu chih-mi* cited by WU Heng-chao above, does not use the term *ch'en-tzu* at all. He does write (3b in the text included in *Tz'u-hua ts'ung-pien, ts'e* 2): "There are many discrepancies between the old scores. Even [texts to] the same tune may have two or three words more or less. Sometimes the division or length of a line varies, having been changed by music-masters, and words have been added or dropped by one singer or another."

The *Ssu-k'u* entry on the book quotes this passage, commenting that it shows that all Sung *tz'u* did not stick to rigid tonal prescriptions, and that Sung singers did not hesitate to "ad lib." In view of this, the *t'i-yao* continues, WAN Shu's statement in *Tz'u-lü* that the *ch'ü* made use of *ch'en-tzu* and the *tz'u* did not, is inconclusive. WU Heng-chao must have assumed from this remark that SHEN I-fu himself had used the term *ch'en-tzu*.

them from others. Once a performer had worked out an effective rendition of a lyric, one which audiences liked, he (or she) would want to keep it on hand. Rival musicians, hearing it, would take it down too, though they might make some changes.[61] Possibly no two groups of musicians played or sang the same tune in exactly the same way; who has ever heard " St. Louis Blues " reproduced note for note as W. C. HANDY wrote it?

Such texts, if they did exist, may even have been for the purpose of preserving the outline of the tune itself, in the absence of any exact system of musical notation. Not much, if anything, is known about T'ang notation, and little enough about that of Sung times.

At any rate, the conclusion drawn from the Chinese comments quoted earlier is that lengthening of *chüeh-chü* lines here and there, and the accretion of additional lines of varying length, resulted from the eventual acceptance—in the minds of both singer and audience—of these accidentals or ornamentations as integral with the song-words. It may be inferred that other singing devices, such as melisma (stretching a syllable over more than one tone, as in HANDEL's " Every valley shall be exa-a-a-a-alted ") sometimes tended to shorten a *chüeh-chü* line by prolonging one word and dropping another.

We can find evidence of some of these processes in *tz'u* texts of the ninth and tenth centuries. Let us begin with song-words consisting of a regular *chüeh-chü* plus interpolated refrains which remain recognizable as such, and the omission of which would leave a self-contained quatrain. In *Hua-chien chi*[62] one of two " Ts'ai-lien tzu " 採蓮子 by HUANG-FU Sung 皇甫松 (late ninth century) goes as follows:

菡萏香連十頃陂 舉棹
小姑貪戲採蓮遲 年少
晚來弄水船頭濕 舉棹
更脫紅裙裹鴨兒 年少

---

[61] The quotation in note 60 from *Yüeh-fu chih-mi*, though written in Sung times, probably is equally applicable to T'ang practice.

[62] *Hua-chien chi chu* 2.12.

Lotus blossoms link their scent for acres along the bank

(Lift oar!)

Where Little Sister, bent on play, takes her time gathering seeds.

(Young folks!)

It's growing late, but she toys in the water, splashing the prow of the boat—

(Lift oar!)

And now she's taken her red skirt off, and wrapped her duckling in it.

(Young folks!)

The other specimen contains the same refrains. The tune to which these poems were written was supposed to have been handed down from Liang times, when princes and emperors wrote words to it.[63] The tune no doubt underwent changes, and many other poets wrote words to it, but none of them before HUANG-FU Sung wrote in these bob-lines. Perhaps they originated in group singing among the people; street musicians may have encouraged their listeners to throw in rhythmic accents, in order to increase their interest and attention and therefore their contributions.

Only the "Lotus Gathering" and "Bamboo Branch" *tz'u* (see n. 59) retained this sort of separable refrain as a convention. These happen to be two of the *tz'u* patterns in which poets nearly always stuck to subject matter that had some connection with the titles, and in these two cases the standard popular refrains also were connected with the subject matter, rather than being meaningless sounds like *i-a-na*. Probably the authors felt that writing in the characters for these refrains gave their song-words a rustic effect. In other *tz'u* patterns (according to the *shih-yü* or *chüeh-chü*-plus explanation), poets replaced meaningless or irrelevant interjections with words adding something to the meaning of a text which might have only the most tenuous connection, or none at all, with the title of the tune. Witness some of the "Yang-liu chih."

All extant T'ang words to this "Willow Branch" tune are straight seven-word *chüeh-chü*.[64] If the shape of the tune was

[63] One of LIANG Wu-ti's "Chiang-nan nung" mentioned on page 9 is subtitled "Ts'ai-lien ch'ü" (cf. *YFSC* 50.1b).

[64] Cf. specimens in *YFSC*, *chüan* 81. The tune was popular in the time of Po Chü-i (see WALEY's *The Life and Times of Po Chü-i* [London, 1949] 196). The earliest extant words to it are nine seven-word quatrains which first appeared in the

such that T'ang singers found it necessary or effective to add a refrain or various rhythmic interjections when performing it, these were not preserved. In the tenth century, however, *tz'u* poets added a three-character phrase after each line. Supposing that these replace with sense what was mere rhythm in T'ang performances, one may analyze a specimen by Ku Hsiung 顧敻 as follows:

1. Seven-word line: 秋夜香閨思寂寥
   a. Interpolation: 漏迢迢
2. Seven-word line: 鴛幃羅幌麝烟銷
   b. Interpolation: 燭光搖
3. Seven-word line: 正憶玉郎遊蕩去
   c. Interpolation: 無尋處
4. Seven-word line: 更聞簾外雨蕭蕭
   d. Interpolation: 滴芭蕉

On an autumn night in her bedroom she broods in the lonely stillness;
Far off the night-watch sounds.
Incense fades among the hangings embroidered with mandarin-ducks,
The flame of her candle flickers.
She's thinking now of her lover off roaming the land,
Wondering where he is,
As she listens to the murmur of the rain outside the blind
Where it drips on the plantain leaves.

As translation readily shows, lines a-b-c-d can be omitted leaving a *chüeh-chü* which makes sense by itself. However, the poet has treated the four three-word lines as integral parts of the song, heightening its langorous sadness not only by their rhythmic monotony, but by what they say.[65]

These *tz'u* from the late ninth and early tenth centuries are

collected works of his friend Liu Yü-hsi 劉禹錫 (*Liu Meng-te wen-chi* 劉夢得 文集 (*SPTK* ed.) 9 (*ts'e* 2).9b-10b). Hu Shih thinks these were genuine popular songs heard, rather than composed, by Liu Yü-hsi. (It seems likely that he at least "doctored" them considerably, as in the case of his "Bamboo Branch" songs—see note 86.) If he also heard refrains or patter-words thrown in between the lines, he did not think them worth recording. Two additional specimens later were inserted in Liu's collection (9.12b) which may or may not be from his hand; at any rate they are quite regular *chüeh-chü*. In very late T'ang, Wen T'ing-yün and Huang-fu Sung were still writing straight *chüeh-chü* under this title.

[65] The actual rhythm, as the poem is read, is 4-3-3 | 4-3-3 | 4-3-3 | 4-3-3.

conveniently simple illustrations of tendencies already operating much earlier. Actually a more advanced stage in the transformation of the *chüeh-chü* is indicated in the "T'iao-hsiao" 調笑 written, or at least written down, by WEI Ying-wu 韋應物 in the late eighth century.[66]

   a. Interpolation: 胡馬胡馬
1. Six-word line: 遠放燕支山下
2. Six-word line: 跑沙跑雪獨嘶
3. Six-word line: 東望西望路迷
   b. Interpolation: 路迷迷路
4. Six-word line: 邊草無窮日暮

Tartar horse, Tartar horse
Loosed afar on slopes of Yen-chih-shan
Paces sand, paces snow, neighs alone,
Looks east, looks west, paths all strange—
Paths all strange, strange paths,
Grass of the marches endless in the sunset.

   This rough translation of what in Chinese is a beautifully evocative, though simple, poem is at least accurate enough to show that the four six-word lines cannot stand alone; without the first "extra" line the poem would have no subject. (The word "interpolation" in the above analysis is used only to suggest an origin for the two four-word lines of the "T'iao-hsiao" pattern.) And a glance at the Chinese shows that without "*lu mi, mi lu*" the last line would have no rhyme.[67] Although it seems to show vestiges of the *chüeh-chü* form, we cannot extract

---

[66] In the Sung edition of his collected works reproduced today the title is written 調嘯詞 (*Wei Chiang-chou chi* 韋江州集 [*SPTK* ed.] 10[*ts'e* 2].5b-6a). The tune is referred to elsewhere by several other names. HU Shih (*op. cit.* 4) says the alternate name "San-t'ai ling" 三臺令 indicates that the tune came from a transformation of the "San-t'ai" air to which quatrains of six-word lines were sung (cf. WEI Ying-wu's two specimens following his "T'iao-hsiao"); that the name "T'iao-hsiao" itself shows that the original words were connected with some kind of game; and that a third name, "Chuan-ying ch'ü" 轉應曲, suggests that variations were made in the words to provide some kind of "answer" song. We do not know whether WEI Ying-wu simply took down the words of these songs as he heard them, or wrote poems of his own modeled on their patterns and possibly their subject matter.

[67] Note that the rhymes are by couplets, regardless of the asymmetry of the rhymed lines.

a *chüeh-chü* from this text. The transformation of that form, under the influence of music, had already reached an advanced stage in this instance.

Meantime the *chüeh-chü* itself, without additional lines, was undergoing some modification. In the poems quoted above it is the quatrain which provides the longer verses, the interpolations which provide the shorter. But many *tz'u* patterns, including some of T'ang date, do not contain any four lines of equal length. Even some of those which are only four lines long fail to qualify, strictly speaking, as *chüeh-chü* because they contain a line or two that is one character short of the prescribed number.

As we have seen, in the seventh and eighth centuries (after the death of Li Po and Tu Fu) the texts which poets wrote for music, or which singers chose for music, had formally correct stanzas; adaptations to make the words fit the music were the business of the musician and no concern of the poet. But from around the beginning of the ninth century a few poets with an absorbing interest in popular music began to take their own liberties with the *chüeh-chü*. Perhaps they were a bit tired of it after all the thousands of correct specimens that had been written. Perhaps they were not conscious of writing altered *chüeh-chü*, but like the musicians, were now thinking in rhythmical terms rather than in terms of literary form. At any rate they produced four-line stanzas like the following:

A. " Yü ko-tzu " 漁歌 by Chang Chih-ho 張志和 (730-810) [68]

    1. 7 words (4/3) 西塞山前白鷺飛
    2. 7 words (4/3) 桃花流水鱖魚肥
    3. 6 words (3/3) 青箬笠。綠蓑衣
    4. 7 words (4/3) 斜風細雨不須歸

---

[68] Cf. *Tz'u-p'u* 1.17a. This is the best known of several " Yü ko-tzu " or " Yü-fu tz'u " 漁父詞 , attributed to a contemporary of Wei Ying-wu, which circulated widely in the Yangtze region. The tune did not outlast the T'ang dynasty; Su Shih, who said Chang Chih-ho's words could not be sung as they stood, added a few characters to the " Yü-fu tz'u " to fit it to the tune of " Huan-ch'i sha " 浣溪沙, and his cousin Li Ju-ch'ih 李如箎 adapted it to the tune of " Che-ku t'ien " 鷓鴣天. Hu Shih (*op. cit.* 8-9) therefore classifies Chang Chih-ho's lyrics with the seven-word *chüeh-chü* song-words of the seventh and eighth centuries, regarding them merely as slightly altered *chüeh-chü* rather than consciously long-and-short verse written to

Before Western Pass Mountain white herons soar,
In the stream beneath the peach-blooms the perch are sleek and fat.[69]
In hat of bamboo leaves and green straw cape
Against the wind and drizzle, one need not go home.

B. "Chang-t'ai liu" 章臺柳 by HAN Hung 韓翃 (fl. 750)[70]

1. 6 words (3/3) 章臺柳。章臺柳
2. 7 words (4/3) 往日青青今在否
3. 7 words (4/3) 縱使長條似舊垂
4. 7 words (4/3) 亦應攀折他人手

Chang-t'ai willow, Chang-t'ai willow,
So green in days gone by, do you grow there still?
Though your long green branches trail as they used to do,
Those that reach and break them off are other hands than mine.

C. "Hua fei hua" 花非花 by Po Chü-i 白居易 (772-846):[71]

1. 6 words (3/3) 花非花。霧非霧
2. 6 words (3/3) 夜半來。天明去
3. 7 words (4/3) 來如春夢不多時
4. 7 words (4/3) 去似朝雲無覓處

The blossoms were not flowers, the vapors were not mists;[72]
She came in the middle of night, and left with the light of day—
Came like a springtime reverie, stayed but a little while,
Went like the clouds of dawn, and was nowhere to be found.

---

conform to the demands of musical phrasing. Two or three other "Yü-fu tz'u" are attributed to CHANG's elder brother Sung-ling 松齡, and about fifteen anonymous ones are thought to be by contemporaries of theirs.

[69] Hsi-sai-shan 西塞山 being in Chekiang, the peach-blossom stream does not indicate the geographical locale of T'AO Ch'ien's famous utopian fantasy (somewhere between Hunan and Never-never land); but the allusion helps to evoke the idyllic rusticity celebrated in that piece.

[70] For the story behind this, cf. Pen-shih shih (TTTS, ts'e 6, 3b-4a). HAN Hung wrote the verses to a woman named Liu 柳 to whose favors men more affluent than he had succeeded. The Chang Terrace was in the suburbs of Ch'ang-an.

[71] Tz'u-p'u 1.15a includes this text as a tz'u-tiao, but nothing is known of any tune connected with it. In Po's collected poems it appears with a group of poems of "emotion" (kan-shang 感傷) in miscellaneous forms (Po Hsiang-shan shih chi 白香山詩集 12.12a). It would be stretching a point to classify it as a chüeh-chü even if it were not one word short, since the tonal pattern is unusually free and the rhymes are in the deflected tone.

[72] I. e., not the flowers and mists of this world.

AOKI observes that such metrical variations as those shown above do not destroy the dominant rhythm established by a seven-word line which, because of the pause at the end of it, or more normally the drawing-out of the final sound, carries eight beats. Quite aside from hypothetical demands of musical phrasing, if the reader will recite to himself the Chinese of the foregoing poems he will find that the six-word lines do indeed fit this eight-beat rhythm, the third and sixth words each accounting for two beats. It is possible, then, that the slightly altered *chüeh-chü*, one or two words short of the standard form, was a purely literary development rather than a musico-literary one. In any event, it was a step toward greater prosodic flexibility and hence toward an easier drawing-together of poetic and musical forms.

AOKI (*op. cit.* 74) illustrates the eight-beat rhythm of both six- and seven-word lines by musical notation, using lines from the " Yü ko-tzu " above:

He contends that in writing or reading Chinese poetry—completely aside from setting it to music—not only six-word lines, but five-word lines as well, are interchangeable with seven-word lines without destroying a basic eight-beat rhythm. The only other normal line being that of four words, the implication is that practically all Chinese poetry is made up of rhythmic combina-

213

tions of four-beat units. AOKI measures as follows a *tz'u* containing lines of three, five, and seven words: [73]

We have here one of Po Chü-i's three *tz'u* entitled " I Chiang-nan " 憶江南, which are translated further on in this essay. Except for WEI Ying-wu's " Tiao-hsiao," which is fairly regular by

---

[73] AOKI (*op. cit.* 75) notates only the first three lines; the last two are notated here on the basis of his conception of the others, except that the last word of the poem is marked with a half-note instead of being broken off with a rest.

AOKI bases his theory of the ubiquitous four-beat unit on the testimony of his own ear when listening to the verbal rhythms of poems and songs read aloud (not sung) by Chinese. It is unexceptionable to anyone who has listened to such reading that the caesura in a six-word line counts for one beat, whether it is realized by melisma or by an inbreathing, and that a pause or hold at the end of any line counts for one beat. The present author, from his limited experience of hearing Chinese poetry read aloud, would not say that the very slight pauses within five- or seven-word lines account for a full beat. In most cases the " punctuation " (讀) in such lines is realized by emphasizing slightly the first word in the next phrase, rather than by making any significant pause. This listener has never heard a two-word unit in regular five-word poetry rendered as AOKI notates *tsui i* (♩ ♩).

In a musical setting, even a Western one, it would certainly be easy enough to interchange five- and seven-word lines. In support of his position that such lines are interchangeable even in reading, AOKI suggests that it explains how " P'ao-ch'iu yüeh " 抛毬樂 (see *Tz'u-p'u* 2.5a), which T'ang poets (e. g., LIU Yü-hsi) wrote in six five-word lines, took the pattern 7-7-7-7-5-7 in the Five Dynasties period. The two versions may well have been sung to the same melody, but the question of purely verbal rhythm is irrelevant. Such songs were never merely " read " aloud; doubtless those who " read " the *Hua-chien chi* to themselves in the tenth century mentally sang each piece, as do those today who read the lyrics of Oscar HAMMERSTEIN II as published in literary form by Simon and Schuster. In most Chinese poetry the rhythmic nature of five-word and seven-word lines appears to be quite different. Although both in the final analysis do contain eight beats, it is submitted that a more general schematization than AOKI's would be:

Seven-word line: ♩ ♩ ♩ ♩ | ♩ ♩ ♩
Five-word line: ♩ ♩ ♩̆ ♩ | ♩   ▬

comparison, these are the earliest real *ch'ang-tuan-chü* to be found among the T'ang *tz'u*.

Whatever the rhythmic basis for the development of such lines from a purely literary standpoint, evidently Po Chü-i was primarily concerned here with reproducing the rhythmic pattern of a popular song. His note following the title " I Chiang-nan " says, " This tune is also called ' Hsieh Ch'iu-niang ' 謝秋娘; each stanza is in five lines." [74] Apparently the poet patterned his line-lengths after a lyric which was already circulating among the singing-girls, musicians, and common people of the South with whom he loved to spend his time.

Although Po Chü-i in his *shih* often referred to the content or atmosphere of popular songs, he does not appear to have said anything about " Hsieh Ch'iu-niang " other than that it was in five lines.[75] Presumably it was the melody which attracted him

[74] *Po Hsiang-shan shih-chi* (SPPY ed.), *hou-chi* 後集. 3 (*ts'e* 7) .12b. WANG Li-ming 王立名 (whose 1703 ed. is the basis for the above) states (*ts'e* 1, *fan-li* 凡例 3b) that notes and comments included with the poems are from the original text unless introduced by the word *an* 按. Variants, which are noted briefly without this character, obviously cannot have been inserted by Po Chü-i, but other comments are assumed to be the poet's own. As WALEY remarks in his biography of Po (p. 217), " his poems, unlike those of LI Po, were not posthumously collected from friends, but were collected and edited by Po himself." The present standard edition of Po Chü-i's complete works, *Po shih Ch'ang-ching chi* (SPTK ed.) omits notes to the poems.

[75] According to *Yüeh-fu tsa-lu* 樂府雜錄 (*TTTS*, *ts'e* 7, 12a), the *t'ai-wei* 太尉 LI [Te-yü] 李 [德裕] (787-849) wrote the original song, while he was military governor of Chih 浙 (Chekiang), in memory of his concubine HSIEH Ch'iu-niang 謝秋娘, the title later being changed from that name to " Meng Chiang-nan " 夢江南. (*Tz'u-p'u*, in the Table of Contents for its first *chüan*, lists the latter under " I Chiang-nan " as one of several alternate titles.) LIU Yün-hsiang (*op. cit.* 126) believes the song was already popular some years earlier. HU Shih (*op. cit.* 7) is also skeptical of *Yüeh-fu tsa-lu*'s accuracy on this score, though he elsewhere (p. 21) remarks that since its author TUAN An-chieh 段安節 was in LI Te-yü's suite in Chekiang around the middle of the ninth century his stories of that period should be dependable. At any rate, present editions of LI Te-yü's works (cf. *Li Wei-kung chi* 李衛公集 [SPTK ed.], *pieh-chi* 別集 4.2a) contain the caption 錦城春事憶江南三首, " Three five-word ' I Chiang-nan ' stanzas about spring in the Brocade City (Ch'eng-tu 成都)," but not the texts of the poems, which have disappeared. Note, however, that the caption indicates that they were in five-word lines, not *ch'ang-tuan-chü*, so that Po Chü-i cannot have been imitating the form of verses written by LI Te-yü. One would further assume from the caption that the latter wrote his three " I Chiang-nan " in Szechuan rather than in Chekiang. If he did write the original

rather than the words. These may well have been lacking in literary merit if they were written by the composer of the tune, or by some prostitute or other entertainer. In either case the concern of the person who devised the words would not have been primarily literary but musical. He or she would not be conscious of whether any given line might resemble part of a *chüeh-chü* or take one of the shapes characteristic of extra phrases thrown in to fill out the musical beat. The lyricist of the wine-shop or brothel would simply make up words to fit the music, paying attention not to how the words would read as a poem, but to how effectively they could be sung. Hu Shih has in mind cases like this when he says that the long-and-short *tz'u* originated not among the poets but " among the people." [76]

Some singing-girls probably did have a certain literary flair and achieved a felicity of expression in their song-words which added to their fame. Unfortunately it is impossible to be sure whether these song-words actually were written by the ladies whose names were often connected with them, or whether the subject matter was in some other manner associated with their private or artistic lives. Po Chü-i several times mentions a Southern *tz'u* by (or about) a woman called Wu Erh-niang 吳二娘.[77] His description of this *tz'u* and quotations from it point to the following text:

---

words to the song, others soon altered them or wrote new ones conforming more closely to the phrasing of the music.

[76] " I suspect that the custom of writing long-short verses following the rhythm of the tunes arose among the people, among the musicians and singing-girls. Literary men were conservative, and kept writing five- and seven-word *shih* as before. But the musicians and singing-girls were interested only in having songs that were good to sing and good to listen to, so they produced long-short verses."—*op. cit.* 15-16.

[77] His *lü-shih* " Chi Yin hsieh-lü " 寄殷協律 (" Sent to the *hsieh-lü* Mr. YIN," *Po Hsiang-shan shih chi* [*SPPY* ed.], *hou-chi*, 8 [*ts'e* 8].8ab) contains the lines 吳娘暮雨瀟瀟曲，自別江南更不聞: " Miss Wu's song of the drizzling evening rain, I hear no more since I left the South." He appends a note that 暮...瀟 is from the words to a tune by Wu Erh-niang of Chiang-nan. Another *lü-shih*, " T'ing t'an ' Hsiang fei yüan ' " 聽彈湘妃怨 (" On hearing ' The Hsiang Maidens' Regrets ' being played," *ibid.* 19[*ts'e* 6].14b), contains the lines 分明曲裏愁雲雨，似道瀟瀟郎不歸: " Clearly in the piece [someone] is grieved by *clouds and rain* [in both the literal and symbolic senses]; it seems to say ' Drizzle, drizzle, he does not return.' " His note says that the words 暮雨瀟...歸 are from a " new Southern *tz'u*."

深畫眉
淺畫眉
蟬鬢鬅鬙雲滿衣
陽臺行雨回
巫山高
巫山低
暮雨瀟瀟郎不歸
空房獨守時

> Brows pencilled deep,
> Brows pencilled pale,
>
> Hair streaming loose, robe girt with cloud,
> The Moving Rain has returned to the Southern Crest.
>
> High on Wu-shan,
> Low on Wu-shan,
>
> Twilight and drizzling rain, and he does not return.
> Alone in the empty house, she passes the time as she can.[78]

This poem, under the title " Ch'ang hsiang-ssu " 長相思, appears in various collections as the composition of Po Chü-i,[79] together with another after the same pattern; but like the two " Ju-meng ling " 如夢令 they are not to be found in his complete works, *Po shih ch'ang-ch'ing chi.*[80] It is evident, however, that he was familiar with this *tz'u* and fond of it. Whoever wrote it, it bears further evidence that popular song-words were already being written in *ch'ang-tuan-chü* in Po Chü-i's time. If he did not directly copy the pattern of the words of Wu Erh-niang's song,

---

[78] For the erotic symbols in the poem cf. the prose " Preface " to the " Kao-t'ang fu " 高唐賦 (*Wen-hsüan* [SPTK ed.] 19.1a-2b, translated by WALEY, *The Temple and Other Poems* [London, 1923] 65-66). In the song, a past erotic experience is contrasted —as usual—with an aftermath of uncertainty and frustration. A courtesan would not have to be highly educated to allude to the " Kao-t'ang fu," which she might never have read; the images in the Preface had long since become standard euphemisms which appear over and over in T'ang song and poetry.

[79] The text here is from *Ch'üan T'ang shih* 32.59b.

[80] Po Chü-i's preface to the last major group included disavows authorship of any poems circulating under his name but not found in *Ch'ang-ch'ing chi.* This preface (71.18b, translated by WALEY in his biography 212) is dated the first day of the fifth month of the first year of Hui-ch'ang 會昌 (845), about fifteen months before his death; he wrote little thereafter, and in sober vein. The " Ch'ang hsiang-ssu " and " Ju-meng ling " *tz'u* are not in the *pieh-chi* or *pu-i* 補遺 sections at the end of WANG Li-ming's edition.

perhaps he found the pattern of " Hsieh Ch'iu-niang " more work-able. On the other hand he may have considered the words of the former admirable as they stood, and have written new words to the latter under the title " I Chiang-nan " because the old ones did not please him. Or perhaps the " Hsieh Ch'iu-niang " tune had already become a customary musical vehicle for celebrating the beauties and joys of Chiang-nan, and Po Chü-i merely wished to add his tribute to China's " Dixieland " in the current manner.

Whether or not he took this experiment in popular-song writing very seriously (his attitude toward his ballads and songs in general hints otherwise), the three " I Chiang-nan " have an evocative magic not wholly dependent on the fact that one comes to them already aware of Po Chü-i's fondness for the South. Per-haps something of their effect will come through in translation:[81]

江南好
風景舊曾諳
日出江花紅勝火
春來江水綠如藍
能不憶江南

It's good to be in the South!
Once I knew well all its sights and sounds:
At dawn the River blossoms redder than flame,
In spring the River waters blue as indigo.
How can I help thinking of the South?

江南憶
最憶是杭州
山寺月中尋桂子
郡亭枕上看湖頭
何日更重遊

Memories of the South:
Oftenest I think of Hang-chou—
Spying out the cassia tree in the moon from the temple in the hills,
Seeing the lake from my pillow in the rest-house.
When shall I ever have such times again?

[81] For the texts below cf. *SPPY, hou-chi,* 3(*ts'e* 7).12b, or *SPTK* 67(*ts'e* 23).12b.

江南憶
其次憶吳宮
吳酒一杯春竹葉
吳娃雙舞醉芙蓉
早晚復相逢

Memories of the South:
Next I remember the mansions of Wu,[82]
A cup of the native wine, leaves of spring bamboo,
A pair of Wu dancers with their wine-flushed flower-faces—
When shall I see them again!

Po's friend LIU Yü-hsi 劉禹錫 later followed this pattern exactly in a delicately wistful little song of his own:

春去也
多謝洛城人
弱柳從風疑舉袂
叢蘭挹露似沾巾
獨坐亦含顰

Spring is gone,
Having paid its respects to us here in Lo-yang.
The willows bending in the breeze seem to be waving goodbye,
Clustered orchises shed their dews as if wetting their handkerchiefs.
And I sit alone, with knitted brow.

LIU captioned this poem " A spring *tz'u* after Lo-t'ien 樂天 (Po Chü-i), to the rhythm of the tune ' I Chiang-nan.' "[83] Po Chü-i had not explicitly stated that he was following his model exactly, but there is no doubt in his friend's case; here we have the first avowed instance of the practice of *t'ien-tz'u*. Note that LIU Yü-hsi's song has nothing to do with the South. Later poets who used this tune (under the alternate titles " Meng 夢 Chiang-nan " and " Wang 望 Chiang-nan ") sometimes celebrated the South (HUANG-FU Sung, LI Yü 李煜) and sometimes used it for other subjects (WEN T'ing-yün 溫庭筠, NIU Ch'iao 牛嶠).

Like WEI Ying-wu[84] of the previous generation, Po and LIU

---

[82] Pre-Ch'in Wu: the Ssu-chou region.

[83] *Liu Meng-te wen chi* (*SPTK* ed.), *wai-chi* 4.14a.

[84] WEI Ying-wu had been Prefect of Soochow when Po Chü-i arrived there around 785 as a boy of thirteen or so, a refugee from the famine at Ch'ang-an. Po wrote

were Northerners whose government service kept them for long periods in the South. All three became fond of the South by making the best of their provincial " exile," savoring the local color of their various stations, visiting the beauty spots of the countryside, and enjoying the talents and company of singing-girls and musicians. All three were connoisseurs of music [85] and sensitive to the particular qualities of the popular music with which they came in contact. Though they sometimes found the local song-words uncouth,[86] they found their patterns worthy of imitation. Because some of the patterns they imitated were ir-regular, they produced the first authenticatable T'ang long-short tz'u, " Tiao-hsiao " and " I Chiang-nan."

many years later that he had been too poor to meet the Prefect socially, but as he grew up he admired WEI's didactic poems and was influenced by them. Cf. the WALEY biography 14, 223-224.

[85] Po Chü-i's and LIU Yü-hsi's musical interests are too well known to require further comment here. YANG Chü-yüan's 楊巨源 Li Mo ch'ui-ti chi 李⊙*吹笛記 (TTTS, ts'e 6, 27.b) associates WEI Ying-wu with music and musicians, and says he was thoroughly versed in musical procedures (洞曉音律).

[86] As in the case of LIU Yü-hsi and the " Bamboo Branch " songs which so fasci-nated him in Szechuan. After describing the village performances with flutes, drums, dancing, and group singing, he concludes: " Of old, when Ch'ü Yüan was living in the region of the Yüan and Hsiang [rivers], the people of those parts summoned the spirits in crude and rustic language; he then wrote the Nine Songs, and even today they sing and dance them in Ch'u. So I also wrote nine ' Bamboo Branch ' tz'u and had trained singers perform them."—Liu Meng-te wen-chi 9 (ts'e 2) .8b. The texts of LIU's nine songs so introduced do not contain the refrains mentioned in note 57.

" The Chinese had in the ninth century," WALEY remarks, " the same complete confidence in the superiority of their own culture that Europeans had in the nineteenth. Liu Yü-hsi found that the shamans of the local aborigines were using in their cere-monies, songs the words of which he considered barbarous and uncouth. He wrote new words in proper literary style, which it is said were used by local singers till long after his time."—Po Chü-i 167.

This does not mean that LIU did not appreciate the native " color " of popular song-words, which he sought to retain insofar as possible in his more polished literary versions. Song-words in more populous regions, such as the cities along the Yangtze, would be more sophisticated, but literary men might still find those by unlettered prostitutes and musicians more touching than tasteful, and undertake to improve them or substitute their own compositions in the same forms. HU Shih (op. cit. 17) thinks this is the way poets began to write tz'u in irregular patterns.

---

* The character is a variant of 謨, written with the radical 言 underneath.

The foregoing pages have suggested some of the influences, literary and musical, which may have prepared the way for the drawing-together of text and melody; but whatever the role and force of these influences, by the end of the first half of the ninth century these three poets had broken the ground for writing poetry directly to music.

Any song-words by Po Chü-i were likely to circulate widely and rapidly; we have his own complaint, in his old age, that his more serious pseudo-odes (the *hsin yüeh-fu*, with their content of social criticism) were neglected while his casual songs and ballads were sung everywhere.[87] Since " Hsieh Ch'iu-niang " or " I Chiang-nan " was already a popular tune, it requires little imagination to suppose that the singing-girls of whom Po Chü-i knew so many, both in his own establishment and elsewhere, took up the words written to it by the famous poet.

One can only conjecture how far the prestige of Po Chü-i stimulated the writing of song-words directly to music regardless of formal irregularities. Since he wrote only three real long-short *tz'u*, all in the same pattern, I hesitate to accord too much influence in this direction even to so famous a name. But the practice undeniably was widespread shortly after his death in 846. Poets often set words to a tune as soon as it was composed. Su O 蘇鶚, who lived only slightly later, reported that when around the middle of the century a barbarian state sent an embassy whose members wore their hair piled up and coiffed with gold like bodhisattvas, actor-musicians devised a tune called " P'u-sa man " 菩薩蠻 (" Bodhisattva Aliens "), and literati repeatedly set words to it.[88] This tune quickly became one of the most popular *tz'u* patterns (7-7-5-5-5-5-5-5). One of the poets who took it up was a young man from T'ai-yüan 太原, who early became famous for his *tz'u* in many forms—including the one used by Po Chü-i in the specimens quoted above. WEN T'ing-yün 溫庭筠 (?820-?870) was the first poet who might be called a *tz'u* specialist; he wrote so many that they filled two collections which circulated separately from his *shih*. The compiler of *Hua-*

---

[87] Letter to YÜAN Chen 元稹 translated by WALEY in *Po Chü-i*, 111 f.

[88] *Tu-yang tsa-pien* 杜陽雜編 (*TTTS*, *ts'e* 1, 31a).

chien chi begins his anthology with some sixty specimens by WEN
T'ing-yün. There is no doubt that he shaped his words directly
to music, for his biography in *Chiu T'ang shu* 舊唐書 says that in
his youth he became noted for just that.

WEN T'ing-yün associated with singing-girls and musicians
more constantly than did Po Chü-i, and under rather different
circumstances. He never became an official, and spent an inor-
dinate amount of his time with other ne'er-do-wells in the wine-
shops and brothels. Perhaps he began writing *tz'u* to improve on
the words he heard sung in such places, and finding his versions
popular with the girls, continued to write for them new words to
old tunes, and sometimes completely new songs. The Preface to
Hua-chien chi [89] characterizes its manner and social atmosphere
as "fanned by the air of the songs of the Northern Lanes
(brothels)," [90] and the very title of the anthology associates its
contents with such a setting.

It is clear then that the forms of WEN T'ing-yün's *tz'u* were
based not on transmutations of literary canons, but directly on
the demands of music. The same may be said of the *tz'u* of his
less prolific contemporary HUANG-FU Sung and of those by WEI
Chuang 韋莊, who lived past the end of the T'ang. Both these
men are also represented in *Hua-chien chi*; of the six patterns
representing HUANG-FU Sung four are *ch'ang-tuan-chü*, as are
all eighteen of WEI Chuang's.

There are a few surviving late T'ang *tz'u* by lesser poets, and
several dozen anonymous ones. Among the latter are the eighteen surviving pieces in the incomplete *Yün-yao-chi tsa-ch'ü-tzu*
雲謠集雜曲子 recovered at Tun-huang in 1907.[91] Their authors

---

[89] *Hua chien chi chu*, "Original Preface" 1.

[90] A later poet who kept similar company, and carried on WEN T'ing-yün's volup-
tuous themes in his poetry, often wrote his *tz'u* on request. "When LIU Yung 柳永
(*chin-shih* 1034) was sitting for his examinations he frequented the brothels. He was
expert at writing song-words, and whenever the *chiao-fang* entertainers hit on a good
tune they would ask him to write a *tz'u* to it before trying it on the public."—YEH
Meng-te 葉夢得 (*chin-shih* 1097), *Pi-shu lu-hua* 避暑錄話 (in *Hsüeh-chin t'ao-
yüan* 學津討源, *ts'e* 32) 2.1b.

[91] These have been edited by Lo Chen-yü 羅振玉 in his *Tun-huang ling-shih*
燉煌零拾, and by CHU Tsu-mou as the first collection in his *Chiang-ts'un ts'ung-shu*

do not appear to have been highly literary, and may well have been entertainers of one kind or another. All eighteen poems are in lines of varying length.

If the principal concern of this discussion has been with the development of *ch'ang-tuan-chü*, it is not because this is the only approach to a study of the origins of the *tz'u*, but because metrical irregularity was the most distinctive literary characteristic of the genre when it began to be generally recognized as a separate branch of poetry. On the other hand, the theory that the *tz'u* is derived from more regular verse forms notably fails to relate it to the *ch'ang-tuan-chü* tradition which is as old as the *Classic of Songs*. Approaching the whole problem of the origins of the *tz'u* from this angle changes the perspective and relegates the aberrant *chüeh-chü* and *lü-shih* verses to the position of modifications brought about under the influence of popular songs, themselves the immediate prototypes of the *tz'u*. I propose to treat this aspect of the history of the *tz'u* in a separate paper.

(see note 56). *Yün-yao chi*, hitherto lost, had been catalogued as containing thirty pieces, but twelve are missing from the book found at Tunhuang (in which the first character of the title is written 云). Small as the collection is, the fact that it is thought to be by various hands and the uncertainty of its date account for the qualifying adjective in the statement at the beginning of this article that *Hua-chien chi* was the first extensive anthology of *tz'u*.

ADDITIONS AND CORRECTIONS

Page 111, line 2: I suggested "tune pattern" as a rendering for *tiao*. "Does 調 in this context," Arthur Waley later wrote to me, "mean anything quite so definite as a tune? Isn't it perhaps more a 'melody-type' and, in that case, rather nearer to the Indian *rāg* than to 'tune' as we use the word when we speak of 'Home Sweet Home' or 'Die Lorelei' as tunes?" Dr. Waley is quite right; I think his comparison with the Indian *rāga*, insofar as I am familiar with them, very apt.

Page 119, note 41, line 14: *For* (see below, p. 00) *read* (see below, p. 143).

Having failed so far to produce the further paper proposed at the end of my article, I hope somebody else will do so. Murakami

Tetsumi 村上哲見 of Koyto University has provided some guide-lines and documentation in his study of Wen T'ing-yün, 溫飛卿 の文學 (*Chūgoku bungakuhō* 5(1956).19–40), and in his 李煜 (No. 16 in the Iwanami series on Chinese poetry, Tokyo, 1955), to which are appended a discussion of and copious translations from the *Hua-chien chi*. Jao Tsung-i 饒宗頤 includes in his 詞籍考, Part I (University of Hong Kong Press, 1963) some suggestive information on early *tz'u*.

A COLLOQUIAL SHORT STORY IN THE NOVEL
CHIN P'ING MEI

by
JOHN L. BISHOP

A COLLOQUIAL SHORT STORY IN THE NOVEL
CHIN P'ING MEI

JOHN L. BISHOP

BOSTON, MASSACHUSETTS

[The substance of this article was first presented as a paper before the fifth annual meeting of The Far Eastern Association on April 1, 1953.]

It is a generally known fact that the novel, *Chin p'ing mei* 金瓶梅 borrows as its point of departure an episode from *Shui hu chuan* 水滸傳 and elaborates the lives and destinies of the characters in that episode. To the best of my knowledge, it has not been pointed out that in chapters 98 and 99 of *Chin p'ing mei* is to be found one of the colloquial short stories contained in the collection *Ku-chin hsiao-shuo* 古今小說, a story entitled " Hsin-ch'iao-shih Han Wu mai ch'un-ch'ing 新橋市韓五賣春情 [" Han Wu Sells Her Love in Newbridge "].[1] Not only do the outlines of the first half of the story appear as a subsidiary episode in the plot of the novel, but both texts show a parallelism of phraseology, extending in passages that vary in length from four or five characters to forty or fifty. The parallelism is interrupted when exigencies of the plot of the novel demand additional or different details from those of the short story, and further variations occur in the names of some of the characters, as well as in orthographic variants and additional colloquial particles. Between the two versions of *Chin p'ing mei*[2] which have come down to us there are still further deviations; but the extent of identical passages and the agreement of narrative details are pronounced enough to convince that here is a case of literary borrowing.[3]

---

[1] [*Ch'üan-hsiang*] *Ku-chin hsiao-shuo* 全像古今小說 (Shanghai, 1947) 3.

[2] a. *Chin p'ing mei tz'u-hua* 金瓶梅詞話, 21 *ts'e*. (Photolithographic reprint of 1617 edition. Shanghai, 1933.)

b. *Tsu-pen Chin p'ing mei* 足本金瓶梅, 16 *ts'e*. (Typeset edition. n. p., n. d.)

[3] Three random examples will serve to demonstrate the degree of similarity among the three texts. No. 1 compares the following passages: (a) *Ku-chin hsiao-shuo* 3.2a; (b) *Chin p'ing mei tz'u-hua* 98.5a; (c) *Tsu-pen Chin p'ing mei* 98.15a. With the

The inevitable question—which borrowed from which?—is not easily answered, for it is impossible to date precisely either the story or the novel. It is true that the collection, *Ku-chin hsiao-shuo* was first printed in 1620 or 1621; [4] but this publication date merely represents the time when its forty stories, previously existing, were issued as an anthology. The editor, publisher and

parenthesized letters representing the same texts, the passages juxtaposed in No. 2 are located as follows: (a) 3.2b; (b) 98.5ab; (c) 98.15b. No. 3: (a) 3.7a; (b) 98.11b; (c) 98.16a. To facilitate alignment of characters, the punctuation used in these editions is here eliminated.

No. 1

(a) 只見屋後河邊泊着兩隻剝船船上　　許多箱籠卓凳家伙四五個
(b) 正臨着　｜｜｜｜｜｜｜｜｜｜｜載着｜｜｜｜｜橇｜活｜｜｜
(c) ｜｜｜　｜｜｜｜｜｜｜｜｜｜｜｜｜｜凳｜｜｜｜

(a) 人盡搬入　　空屋裏來
(b) ｜｜｜｜樓下｜｜　｜祚裏｜
(c) ｜｜｜｜｜｜｜｜｜｜｜

No. 2

(a) 吳山問　主管道　甚麼人　不問事由擅自搬入我屋　來
(b) 經濟｜謝｜｜　是｜｜｜　｜｜自｜｜｜｜｜｜祚裏｜
(c) 敬｜｜｜｜｜　｜｜｜也｜｜一聲｜｜｜｜｜｜｜

No. 3

(a) 自別尊顏思慕之心未嘗少怠懸懸不忘於心向彖期約妾倚門凝
(b) ｜｜｜｜｜｜｜｜｜｜｜｜｜｜｜｜｜｜｜｜于｜｜｜｜｜｜｜｜
(c) ｜｜｜｜｜｜｜｜｜｜｜｜｜｜｜｜｜｜｜｜｜｜｜｜｜｜

(a) 望不見降臨　　昨遣八老探拜　　不遇而囘妾移居在此甚是荒
(b) ｜｜｜｜｜蓬華｜｜｜｜｜問起居｜｜｜｜
(c) ｜｜｜｜｜｜｜｜｜｜｜｜｜｜｜｜｜｜

(a) 涼聽聞貴恙灸火疼痛使妾　　　坐臥不安
(b) ｜｜｜｜欠安　今｜空懷悵望｜｜悶懨
(c) 聞知｜｜｜｜　｜｜｜｜恨｜｜｜｜

[4] For a description of this and other early editions cf. SUN K'ai-ti 孫楷第, *Chung-kuo t'ung-su hsiao-shuo shu-mu* 中國通俗小說書目 (Peip'ing, 1932) 122-8; *So-chien Chung-kuo hsiao-shuo shu-mu t'i-yao* 所見中國小說書目提要 (Peip'ing, 1933) 17.

owner of the library from which the originals were taken has been identified as FENG Meng-lung 馮夢龍, editor and author of many popular works at the end of the Ming dynasty.[5] Other stories in this collection are to be found in prior compilations dating from at least early Ming times.[6] These versions in turn are believed to have been based on *hua-pen* 話本 or printed versions of the prompt-books used by popular story-tellers, written versions of oral stories which began to be printed sometime in the late Sung or early Yüan period. It is very possible that the particular story with which we are concerned has a similar history, although no earlier version of it has come to light. It might, in short, have been narrated orally and committed to written form at any time between the eleventh century and 1621.

Chin p'ing mei presents a dating problem of a different sort. Obviously written by a single author, it shows no evidence, vestigial or conventional, of an oral tradition of story-telling as do other *hsiao-shuo* of the period. But the question is who was its author. The earliest mention of the work in the *Shang-cheng* 觴政[7] of YÜAN Hung-tao 袁宏道 places the novel before YÜAN's death in 1610. A more informative reference from a slightly later work is to be found in *Ku ch'ü tsa yen* 顧曲雜言. The author, SHEN Te-fu 沈德符, relates that in 1606 he borrowed a

[5] SHIONOYA On 鹽谷溫, " Kuan-yü Ming-tai hsiao-shuo *San-yen*" 關于明代小說『三言』 (Chinese translation in WANG Fu-ch'üan 汪馥泉, *Chung-kuo wen-hsüeh yen-chiu i-ts'ung* 中國文學研究譯叢, Shanghai, 1930) 23-24. For biographical facts and a bibliography of works cf. JUNG Chao-tsu 容肇祖, " Ming Feng Meng-lung ti sheng-p'ing chi ch'i chu-shu " 明馮夢龍的生平及其著述, *Lingnan hsüeh-pao* 嶺南學報 2.2 (1931).61-91.

[6] *Ch'ing p'ing shan t'ang hua-pen* 清平山堂話本 compiled by HUNG P'ien 洪楩 during the Chia-ching 嘉靖 period (1522-1566). A portion discovered in Japan has been issued in facsimile reprint, Tōkyō, 1928 and Peip'ing, 1929. Another portion recovered in China has been reprinted in facsimile under the title, *Yü-ch'uang I-chen chi* 雨窗欹枕集, Peip'ing, 1934.

Ching-pen t'ung-su hsiao-shuo 京本通俗小說 (reprinted as *Sung-jen hsiao-shuo* 宋人小說, Shanghai, 1940) was long considered a compilation of Yüan date, but for recent doubts on this dating cf. YOSHIKAWA Kōjirō 吉川幸次郎, " Shijō Chō shukan hyō " 『志誠張主管』評, in his *Chūgoku sambun ron* 中國散文論 (Tōkyō, 1949) 190-220; and Jaroslav PRŮŠEK, " Popular Novels in the Collection of Ch'ien Tseng," *Archiv Orientální* 10 (1938).292-3.

[7] *Pao yen t'ang pi-chi* 寶顏堂祕笈 (Shanghai, 1922) (*hsü-chi* 續集 ts'e 8) 2b.

complete copy of the novel from YÜAN Hung-tao's younger brother and made a transcript for himself, which later in Suchow he showed to his friend FENG Meng-lung. The latter, delighted with the novel, urged a printer to buy it; but SHEN piously refusing to take the moral responsibility for propagating such an immoral work, locked up his copy. It is not clear whether he relented or whether other copies were circulating, but he adds that a short time later the novel was to be bought all over Suchow. In conclusion he reports having heard that *Chin p'ing mei* was written by a prominent scholar of the Chia-ching 嘉靖 period (1522-1566) and that it satirizes actual events and personages of that era.[8]

From this reference it is clear that the authorship of *Chin p'ing mei* was already a mystery a hundred years after its supposed time of composition. On the basis of the last hint by SHEN Te-fu, several authors have been suggested, including FENG Meng-lung himself, but none has been substantiated. Therefore, returning to our two texts, external evidence yields nothing to settle the precedence of either, beyond the inconclusive fact that the compiler of *Ku-chin hsiao-shuo*, FENG Meng-lung, had read *Chin p'ing mei* some ten years before his story collection was printed.

An immediate supposition is that FENG may have abstracted a portion of the novel which he had found entertaining to form one of the stories in his anthology. It is true that one story in a subsequent collection[9] is known to be his work and others show evidence of restyling by him;[10] but internal evidence strongly supports the view that FENG could not have composed this particular *hsiao-shuo* from incidents in *Chin p'ing mei*. In fact, such

---

[8] *Ku-ch'ü tsa-yen* (*Sung-fen shih ts'ung-k'an* 誦芬室叢刊) (*ts'e* 74) 14b-15a. I am indebted to Professor YANG Lien-sheng for calling my attention to a further substantiation of this dating, the fact that of the eighty-eight songs (*ch'ü-tzu* 曲子) in the novel, sixty appear in *Yung-hsi yüeh-fu* 熙樂府 and forty-six in *Tz'u-lin chai-yen* 詞林摘豔, both compilations of the Chia-ching period. Cf. FENG Yüan-chün 馮沅君 *Ku-chü shuo-hui* 古劇說彙 (Shanghai, 1947) 191-5.

[9] "Lao men-sheng san shih pao-en" 老門生三世報恩, *Ching-shih t'ung-yen* 警世通言 8.

[10] Compare, for example, from *Ch'ing p'ing shan t'ang hua-pen* the stories "Li Yüan Rescues a Red Snake in Wu-chiang" (李元吳江救朱蛇) and "The Ring" (戒指兒記) with *Ku-chin hsiao-shuo* 34 and 4.

evidence points to the conclusion that " Han Wu Sells Her Love in Newbridge " is one of that group of stories which stems from narration by marketplace story-tellers.

Structurally the narrative shows its early origins in a formal introduction, consisting of five short sketches of past rulers who were addicted to feminine beauty, each sketch prefaced or concluded by a poem in seven-word lines. Two of the latter are *yung-shih* 詠史 poems by Hu Tseng 胡曾 (c. 860) [11] and another is a quotation from Po Chü-i, 白居易, " Ch'ang hen ko " 長恨歌 (" Song of the Everlasting Remorse ") . [12] The historical anecdote in prose connected with each poem serves as a commentary to clarify the allusions in the verse. This entire introduction, with the exception of the poem and sketch devoted to Duke Ling of Ch'en 陳靈公, parallels a passage in *Hsüan-ho i-shih* 宣和遺事 [13] where the identical poems and the same anecdotes in slightly different phrasing occur juxtaposed in the same sequence. The narrative method of these two examples conforms closely to the conventional pattern of alternating cryptic poem and explanatory narrative prose to be found in such a work of early vernacular literature as *Ch'üan-hsiang p'ing-hua* 全相平話 which exists in a Yüan printing. [14] Used to introduce a main story, it becomes a characteristic device in early *hua-pen* and one which presumably allowed the narrator to go on improvising historical parallels to the tale which was to follow until his audience reached profitable proportions and at the same time hold the attention of early-comers.

The remainder of the story contains an unusual number of intrusions by the narrator in the form of rhetorical questions and of colloquies between narrator and audience. If such rhetorical

[11] *Hsin tiao chu Hu Tseng yung-shih shih* 新彫注胡曾詠史詩 (*Ssu-pu ts'ung-k'an* ed.) 2.9b-10a; 2.10b.

[12] *Po Hsiang-shan shih-chi* 白香山詩集 (Shanghai, 1915) 12.6a.

[13] *Hsüan-ho i-shih* (*Ssu-pu pei-yao* ed.) (*ch'ien chi* 前集) 2a-3a.

[14] Reprinted in facsimile by the Commercial Press, Shanghai, n. d. For a discussion of the use of *yung-shih* poems, especially those of Hu Tseng, in early *p'ing-hua*, cf. Chang Cheng-lang 張政烺 " Chiang-shih yü yung-shih shih " 講史與詠史詩 *CYYY* 10 (1948) .601-645 and J. I. Crump, Jr., " *P'ing-huà* and the Early History of the Sān-kuó Chìh," *JAOS* 71 (1951) .249-256.

devices are a survival from an earlier time of oral presentation, they have come to be retained in a written version as literary conventions; but in this particular story their extended use and spontaneous quality differ markedly from their unimaginative use as stereotyped formulae in *hsiao-shuo* of later composition. In addition, the story is set in and around Linan, the Southern Sung capital where the oral tradition of story-telling flourished most extensively,[15] and many specific references are made to local streets, gates, markets and nearby villages, references which assume an audience's familiarity with this localized topography.

The text of the story as preserved in *Ku-chin hsiao-shuo* contains some unintelligible passages which can only be regarded as corrupted text.[16] It is inconceivable that FENG Meng-lung could have written such garbled prose, and its presence in his collection of stories even argues against any thorough editing on his part. In the matter of diction, the story version contains peculiar expressions, the meaning of which must be determined from context or, in some instances, by recourse to similar colloquialisms in *Shui hu chuan* or the Yüan dramas. Where such expressions occur in common passages, the *Chin p'ing mei* text avoids them, using either a more familiar term or an expanded phrase, the meaning of which is readily apparent.[17] If these expressions are examples of colloquial idiom which by Ming times had become obsolete, it would be natural for the author of *Chin p'ing mei* to substitute more up-to-date locutions. It should also be noted that in parallel passages the text of the novel generally uses a more vernacular style with a greater number of colloquial particles than does the short story.

The plot as a whole confirms the supposition that the story is an old one. Here is a combination tale of passion and retribution through a supernatural agency, specifically a Buddhist

---

[15] The principal account of the prevalence and variety of popular narrators in thirteenth century Linan is NAI Te-weng 耐得翁, *Tu-ch'eng chi-sheng* 都城記勝 (*Lien-t'ing shih-erh chung* 棟亭十二種) (*ts'e* 1).10a-b.

[16] Cf. *Ku-chin hsiao-shuo* 3.4a, lines 11-12.

[17] For example:

無用閑闊 (*KCHS* 3.4a)— 靠老婆衣飰肥宅 (*CPMTH* 98.9b)

坐在街簷石上 (*KCHS* 3.7b)— 坐在沿街石臺基上 (*CPMTH* 98.10b).

231

one. In brief, Wu Shan, a well-to-do young man, having fallen
victim to the mercenary scheming of a young prostitute named
Chin-nü, is possessed by the ghost of a Buddhist monk who has
died under similar circumstances in the same house. Possession
takes the form of a violent illness by which the ghost intends to
bring Wu Shan to hell to keep him company. Only by a last-
minute confession on the part of the renegade and the appease-
ment of the ghost by Wu Shan's father, is he saved to enjoy the
more respectable pleasures of repentance and moralizing. The
strong Buddhist element as well as the juxtaposition of highly
erotic material with an avowed moral purpose are characteristic
of a group of *hsiao-shuo* which in other respects, such as structure,
diction and setting, show evidence of early origins. The plot as
a whole has structural unity, the early half dealing with the
seduction by Chin-nü and the latter half concerned with the
wages of sin, both forming a well-integrated and homogeneous
narrative.

Turning the light of such internal evidence on the *Chin p'ing mei*
version, we notice at once that only the first part of the plot is
used. The supernatural agent of retribution has been eliminated
or has been ingeniously reduced to an episode in which a certain
local bully, " Tiger Liu," who is very much alive, threatens the
life of the hero, Ching-chi. The suppression of the supernatural
element which is itself characteristic of the colloquial *hsiao-shuo*,
argues for the theory that the realistic author of *Chin p'ing mei*
is the one who is adapting. Of course, all rhetorical interpolations
reminiscent of oral presentation are absent in this novel intended
to be read.

In a comparison of the two versions designed to catch a narrative inconsistency in one which does not appear inconsistent in
the other, the *Chin p'ing mei* version yielded two definite examples.
The first occurs when Ching-chi is compromised by the girl. A
prominent incident of the scene is the theft of the hero's gold
hairpin, by which ruse the girl lures Ching-chi upstairs to her
room. Since he is all too willing to comply with her plan, the
device seems somewhat pointless. In the short-story version,
however, Wu Shan, disgusted with the girl's behavior, is reluctant

to consent to her designs; and her theft of his hairpin is functional as a means of getting him into a compromising position against his will.

The second inconsistency is a more definite one. After the previous scene a long interval elapses before the lovers' next meeting. In the short story the reason for the delay is Wu Shan's ghostly illness. In the novel it is the suspicion of the hero's wife that her husband is philandering and her restricting his liberty. But illness is used as a pretext by Ching-chi to explain to the girl why he cannot come to her. Thereafter, however, in the novel this pretended illness is treated as the reality which it is in the other version. For example, when Ching-chi receives gifts from the prostitute, he explains them away as gifts from the manager of his wine shop who had heard he was ill. His wife believes this fiction, a fact which is odd, knowing as she must that he has not been ill. When he does go out, another member of his household warns him to take a sedan chair lest he overtire himself, indirectly referring again to the non-existent illness. But in the short story, since Wu Shan has been ill in bed for two weeks, it is natural that his wife should believe the lie about the gifts and the solicitude that he not over-exert himself is explicable.

On the strength of internal evidence alone, therefore, it seems certain that the author of *Chin p'ing mei* has not only taken his prologue from *Shui hu chuan*, but has incorporated into two of his final chapters a colloquial *hsiao-shuo* current in his day. The version to which he had access may not have been identical with that now found in *Ku-chin hsiao-shuo*, but the similarity of their common passages strongly favors the supposition that it was.

Whatever value such a discovery may have lies in the light it can shed on the sudden appearance of a novel which in many respects was an innovation in Chinese fiction. While a series of prototypes can account for the final versions of *Shui hu chuan* and *San-kuo chih yen-i*, such a realistic narrative of everyday life as *Chin p'ing mei* has been regarded as an isolated and spontaneous creation. The presence in it of a colloquial short story suggests that precedents for the use of sublunary rather than legendary fictional material and for a realistic rather than formulistic narra-

233

tive technique exist in the large body of *hsiao-shuo,* only a portion of which has probably survived in the three collections edited by FENG Meng-lung. In a certain group of love stories within that genre are to be found narrative techniques which depend upon naturalistic dialogue, use of accurately observed details of domestic life, and attention to the narrative logic of cause and effect in plotting, techniques which the unknown author of *Chin p'ing mei* has employed with a greater degree of sophistication and effectiveness.

Rather than stress too much the assertion of " literary borrowing " with its implied accusation of artistic poverty, we might better reemphasize the fact that there existed in Chinese vernacular literature a common fund of narrative materials: tales, stories, and historical anecdotes, upon which dramatist, story-teller and novelist alike might draw without guilt of plagarism. That the author of *Chin p'ing mei* followed this established precedent does not detract from a truly great creation. Rather one must admire the ingenuity and fine craftsmanship with which he has incorporated traditional material into the design of his essentially original novel and has left scarcely a joint to be discovered. If we were to feel that the artistic stature of *Chin p'ing mei* was lessened by such a discovery, we would also be forced to lower our esteem for such works as *The Canterbury Tales* and many of SHAKESPEARE's plays on the same grounds.

ADDITIONS AND CORRECTIONS

Page 397, note 8, line 4: *After Yung-hsi yüeh-fu supply* 雍

SOME LIMITATIONS OF CHINESE FICTION

by

JOHN L. BISHOP

SOME LIMITATIONS OF CHINESE FICTION

JOHN L. BISHOP
Harvard University

One wonders what the general reading public has made of the translations of traditional Chinese fiction which have recently appeared in bookstores, in several instances in paper-bound series usually devoted to up-to-date novels of violence and vampires. Chinese colloquial fiction before the coming of Western influences certainly contains enough of both murder and adultery to give the average reader a sense of literary familiarity; but the thoughtful reader must be puzzled by an undefinable inadequacy, by a feeling of literary promise unfulfilled, to which even the student of Chinese stories and novels must confess. Unconsciously conditioned as are we all to the premises and achievements of European fiction, we cannot fail to weigh this fiction of another culture in the same balance and find it vaguely wanting. In the following pages I intend to isolate several of the factors which contribute to our impression of disappointment upon reading those works which have long been a source of delight to the Chinese.

In doing so, I must admit to taking arbitrarily the fiction of the West as a standard against which to measure works in a wholly unrelated literature, a questionable procedure if used merely to arrive at a value judgment, but a justifiable method if used to localize and appraise the different development in comparable genres of two distinct literatures. Western fiction, moreover, has always displayed a vitality which makes it an eminent criterion, a vitality which has led to capacity for experimentation, variation and theorizing extending down to the present day. Not for a century at least has the conviction prevailed in the West that "a novel is a novel, as a pudding is a pudding, and that our only business with it could be to swallow it." Once emancipated from the stigma of immorality, European fiction gradually became recognized as a justifiable form of truth, as valid as that offered by the historian, the philosopher or the painter. Henry James's defense of the art of fiction merely served notice on a state of affairs already brought about by the work of Jane Austen, George Eliot and Dickens.

With the recognition of fiction as a form of truth, the reader's concept of fiction shifted from the romance and the tale to the novel and the story. He now expects, however unconsciously, in such literary forms a writer's personal, consistent view of life, and he expects as a concomitant, a personal and consistent literary style. In other words, he takes for granted on the lowest level of his reading some degree of literary realism, some accuracy in the description of the behavior of human beings as individuals in conformity or in conflict with a plausible social environment, and on the higher levels of his reading he looks for some degree of philosophical realism by which the author's personal judgment of a social condition or of a human problem is made clear. The reader of fiction, then, unless he confines himself to the watered-down and sugar-coated imitations which inevitably swarm in the wake of those original works that appear like revelations, wishes not only to be diverted by "lies like truth" but to be edified by a personal vision of truth seen through the medium of lies or fictions. If we accept this concept of fiction, it seems to me that the traditional colloquial fiction of China is limited in two respects: the one a limitation of narrative convention, the other a limitation of purpose.

Perhaps one should not distinguish so boldly between style and content, between form and function. Recent criticism insists that what is to be said inevitably shapes the manner in which it is said. But the point I wish to make is that in the Chinese fiction we are discussing this integration is imperfect, that in fact, primitive narrative conventions were retained long after the narratives had begun to change in scope and purpose, and that new themes were forced into old molds to the detriment of the final product.

These primitive conventions stem from an earlier period when colloquial fiction was part of an oral tradition of literature. From scattered references in T'ang and Sung sources[1] and from descriptions of the two capital cities of the Sung dynasty,[2] it is clear that the marketplace storyteller was a common social institution with well-established traditions in those periods, if indeed, he has not always been a feature of Chinese urban society. Faced with the problem of entertaining an audience, constantly coming and going, illiterate yet shrewd as are those who rely on the evidence of things seen and heard, rather than read, he evolved in the course of time a narrative genre which solved that problem.

Drawing for his materials upon historical records, Buddhist and Taoist hagiographies, tales in the literary language, and even celebrated local scandals, he was guided by at least one common criterion,—sensationalism, either supernatural, murderous or sexual. These materials he elaborated, giving to the terse originals a wealth of naturalistic, but nonetheless fanciful detail, calculated to convince his auditors of the plausibility of what was inherently incredible.

The form in which these stories were presented to the listening public had several characteristic features. They were introduced by a prologue in which anecdotes and poems related to the theme of the main story were strung out until the audience reached profitable proportions. Poetry was frequently intro-

---

[1] Tuan Ch'eng-shih[e], *Yu-yang tsa-tsu[f] hsü-chi* [*Supplement to the Yu-yang miscellanea*], 4: 11a in *ts'e* 56 of *Hupei hsien cheng i-shu[g]*; Su Shih[h], *Tung-p'o chih-lin[i]* [*Literary remains of Tung-p'o*], 6.

[2] The principal descriptions are: Meng Yüan-lao[j], *Tung-ching meng-hua lu[k]* [*Memories of the eastern capital*], 7b in *ts'e* 3 of *T'ang-Sung ts'ung-shu[l]*; Nai Te-weng[m], *Tu-ch'eng chi-sheng[n]* [*The Wonders of the capital*], 10a–b in *ts'e* 1 of *Lien-t'ing shih erh chung[o]*; Chou Mi[p], *Wu-lin chiu-shih[q]* [*Hangchow that was*], 6: 11a–12b, in *ts'e* 250–2 of *Pi-chi hsiao-shuo ta-kuan[r]*.

duced into the recital, probably with musical accompaniment.[3] Originally such verses may have had an integral function in the story; later they served as a commentary, a verification, a means of delaying a climax, or merely as an embellishment. The narrator felt free to intrude in his own person into the story, lecturing his auditors on some moral problem raised by the plot, answering questions which he assumed to be in their minds, even exhibiting to them some tour de force of narrative logic which they might have missed. Characters in the story were often made to recapitulate the plot for the benefit of late-comers in the audience. The narrative style relied greatly on the use of dialogue to advance the plot; and presumably such a style allowed the storyteller to differentiate speakers in a semi-dramatic fashion. Also close to theatrical technique is the manner in which the movements of characters are meticulously described, so that by recording their sittings, risings, bowings and the like, we retain a constant and clear picture of the scene. Lastly, the stories are limited in length to what the attention span of a listening audience might comfortably endure. In the few surviving examples where we have stories which were presented in two installments or *hui*[a], the break occurs at a point of high suspense in the plot and thus ensures the return of the audience at the next session. In general, these stories betray the narrator's concern with using such conventions with the highest degree of craftsmanship rather than any interest on his part in adapting the old or inventing new narrative devices to fit some particular story.

At this point I must confess that there are no verbatim recordings of a Sung storyteller's recital. The characteristics of his style just enumerated have been drawn from later written versions of his stories, versions which appeared in printed collections during the Ming period[4] but which unquestionably existed in written form in Yüan and Sung times. What is of interest is the fact that during these centuries of development from an oral to a written genre, the oral conventions persisted to such a degree in versions designed to be read. With the conservatism characteristic of Chinese literature, these once functional literary devices have been retained as unessential literary clichés. As a sort of author's commentary on the story he is relating, their cumulative effect is to destroy the illusion of veracity which naturalistic plot details attempt to create; and the retention of such conventions has impeded the development of a realistic narrative technique toward its ultimate goal of producing an effect of actuality.

---

[3] Many of the poems used are *tz'u*[s], a form originally associated with musical accompaniment. The narrator's cues to his accompanist before each poem have still survived in the text of one story, *Ching-shih t'ung-yen*[t], 38.

[4] *Ch'ing p'ing shan t'ang hua-pen*[u] [*Colloquial stories from the Ch'ing p'ing shan studio*] was compiled by Hung P'ien[v] between 1522 and 1566; *Ching-pen t'ung-su hsiao-shuo*[w] [*The capital edition of colloquial stories*] is of disputed compilation date but contains materials antedating the Ming period; *Ku-chin hsiao-shuo*[x] [*Stories old and new*] with an alternate title *Yü-shih ming-yen*[y] [*Clear words to instruct the world*]; *Ching-shih t'ung-yen* [*General words to admonish the world*]; and *Hsing-shih heng-yen*[z] [*Constant words to arouse the world*] were edited and published as a series by Feng Meng-lung[aa] in 1621, 1625, and 1628 and are referred to collectively as the *San-yen*[ab] [*The three yen*].

Fielding's digressions on the prose epic in *Tom Jones* and Thackeray's more intimate intrusions into his novels have called forth similar criticism in the West.

So far I have spoken only of colloquial short stories as they have been preserved in the *San-yen* and other anthologies of the Ming period. While this genre was developing during the Sung and Yüan periods, the colloquial novel was also evolving by a process of accretion from groups of such short narratives, possibly combined with dramatic versions dealing with a common pseudo-historical episode.[5] But the survival of conventions used by oral narrators is still evident in these novels. Prose is mixed with verse and dialogue is used extensively. Chapters, still called *hui*, usually end at a climax; and the reader is urged in a stereotyped formula to hear what happens in the next installment.

Probably the most notable influence of its early origins on the novel and one most disturbing to the Western reader is the heterogeneous and episodic quality of plot. In *Shui-hu-chuan*[6] he is expected to follow a story involving 108 heroes, over a third of whom have a major role, and in *San-kuo chih yen-i*[7] he must cope with the shifting fortunes and myriad adventures of the rulers and military leaders of three warring states. Not since the Arthurian romances and Malory have Western readers been entertained with such a plethora of characters and incidents within the confines of a single literary work. These accretive novels, then, retain the meticulous narrative style of their original materials, a style which is preoccupied with surface reality, presenting to the reader a clear visual picture of outward appearance and movement and a verbatim account of dialogue. In addition, the structure of their plots is marked by episodic variety, bound by a tenuous unity of historical or pseudo-historical theme.

In the subsequent fiction of the Ming period, writers of novels and short stories accepted the narrative conventions of the *San-yen* collections and the *Shui-hu-chuan*. Aside from an inherent literary conservatism, they probably had an added motive in doing so: the need for the literatus to conceal any connection with the vulgar literature. In an atmosphere where fiction in the colloquial language was considered almost a defilement of the long-treasured and esoteric art of writing, few members of the scholarly élite could risk being known as compilers of a version of popular fiction or as authors of a new specimen in any of its genres. Use of the collective, traditional style of the storyteller, therefore, served as an excellent means of preserving anonymity.

The result of this fact is, to the Western reader, a curious absence of personality in the style of such fiction, a monotonous preoccupation with "story" rather than with an individual mode of telling the story. Chinese fiction for this reason has no Cervantes, no Richardson, no Jane Austen, who, relying to be sure on the work of predecessors, nevertheless gave the literary forms they found at

---

[5] A thorough study of the development of one such novel is Richard G. Irwin's *The Evolution of a Chinese Novel: Shui-hu-chuan* (Cambridge, Mass., 1953).

[6] Translations: J. H. Jackson, *Water Margin* (abridged) (London, 1937); Pearl Buck, *All Men Are Brothers* (New York, 1937).

[7] Translation: C. H. Brewitt-Taylor, *San Kuo or Romance of the Three Kingdoms* (Shanghai, 1925).

hand a turn in a new direction or added new depth of insight, even a new dimension to be afterwards associated with that form.

Chin p'ing mei,[8] a novel written in the sixteenth century which is for the most part an original production by a single hand, illustrates this limitation. Its unknown author has taken a narrative form common among traditional short stories, the exemplary tale in which the ultimate penalties of a life of dissipation are presented by graphic illustration. He has, however, expanded this theme by tracing the spread of moral laxity within a large family unit and ultimately among more distant family connections. The originality of his work lies not so much in the novelty of this theme as in the magnitude of its illustration. By extending moral retribution beyond the limit of the individual sinner to the family and, by implication, to society as a whole, he has introduced an innovation into the genre of fiction. While to a limited extent the innovation is prepared for by the large canvas of social forces at work which *Shui-hu-chuan* presents, the earlier novel does not anticipate the masterly depiction of domestic life in all its complicated detail which is one of the principal attractions of *Chin p'ing mei*.

In style and narrative technique, on the other hand, *Chin p'ing mei* is indistinguishable from the fiction which precedes it. It continues quite naturally and without a noticeable variation in style from the *Shui-hu-chuan* incident which is its point of departure. Furthermore, a colloquial short story embedded in one of its later chapters[9] is stylistically indistinguishable from the context in which it appears. In other words, *Chin p'ing mei* employs most of the inept narrative conventions of earlier fiction, except obvious intrusions by the narrator, and binds together a wealth of loosely related episodes, giving these a degree of homogeneity by its implicit unity of theme. To the Western reader its final effect of satiety with the carnal life is a result of an overwhelming accumulation of incident rather than of the careful selection of telling narrative details. Yet it is a testimonial to the fine craftsmanship of traditional narrative technique that, despite its wooden and impersonal style, the novel carries a high degree of conviction in its details and an irresistible impact in its entirety.

Multiplicity of detail, striving to reproduce the social macrocosm rather than to explore the human microcosm, appears to be a characteristic of Chinese fiction inherited from the accretive methods by which its prototypes evolved. It is the rare novel in Western literature—*War and Peace*, Proust, and Dos Passos come to mind—which attempts the panoramic social picture of *Shui-hu-chuan*, *Chin p'ing mei* or *Ju-lin wai-shih*.

Another characteristic of Chinese fiction disturbing to the Western reader is the mingling of naturalism and supernaturalism within the same narrative. Poe's requirement of unity of effect or impression within a single narrative has long prevailed in Western literature. If a novel or story is to be a fantasy, the reader demands to know this from the start and to have the tone of fantasy maintained

---

[8] Translations: Clement Egerton, *The Golden Lotus* (London, 1939); Bernard Miall, *Chin Ping Mei, The Adventurous History of Hsi Men and his Six Wives* (abridged) (New York, 1938).

[9] See John L. Bishop, "A Colloquial Short Story in the Novel *Chin p'ing mei*," *HJAS*, XVII (Dec. 1954), 394–402.

consistently. His willing suspension of disbelief varies greatly in degree, if not in quality, when reading for example *Vatek* and *Vanity Fair*. The intervention of ghosts or deities into a perfectly mundane sequence of events disturbs not only his sense of illusion but his standard of literary propriety. When, on the other hand, he opens *Hsi-yu-chi*[10] and begins to read of a rock which became pregnant and gave birth to a stone monkey, he is prepared to accept all of the delightful fantasies which follow.

It may be argued that to the society for which these fictions were written, there was no incongruity in the mingling of flesh and blood with ghosts and gods, and hence no violation of plausibility. But it is apparent that at least by Ming times a definite rationalism had begun to make its appearance in fiction. While the legendary story of exemplary behavior and supernatural marvels continues to be repeated, it is the love story or erotic narrative with a domestic setting in which creative effort is centered. Even in the material of the *San-yen* collections we can observe the process by which the love element in some tale of wonder has begun to be expanded with realistic detail until its length is out of all proportion to that of the matrix story and its naturalistic style out of keeping with its original context.[11] *Chin p'ing mei* is an excellent example of this trend. Except at the end where the visions conjured up by the mysterious Buddhist priest point up the moral significance of a seemingly immoral story and except for the appearance of Wu Ta's ghost retained from *Shui-hu-chuan*, all marvels have been carefully suppressed, even where borrowed materials suggest their use.

The tendency toward rationalism in fiction can be seen very clearly in the anonymous preface to *Chin-ku ch'i-kuan*[b] [*Wonders old and new*], a late Ming anthology of stories selected from the *San-yen*. The author of the preface felt that some explanation of the character *ch'i* for "strange" or "wonderful" in the title was needed. "Strange" to him means not the impossible but the unusual, not that which violates natural or human principles, but that which on rare occasions exactly conforms to them. Marvels to him are those paragons of constancy to the cardinal human virtues recorded in history and romanticized in popular fiction. In the preface he attempts to shift the focus of attention from the miracles and ghostly visitations in many of the stories to the exemplary heroes of a few of them. But his rationalization also implies that the newer, more naturalistic love story, recording as it does striking lapses from exemplary behavior, has in this type of "strangeness" a moral value.

But Chinese fiction, while partially developing a naturalistic method, never wholly accepts its obvious concomitant, a naturalistic and purely human view of life. Always behind the plausible interplay of human emotions, human acts and consequences, lies the assumption of supernal forces directing the ultimate fate of the characters. In the best of the naturalistic stories a high degree of coincidence has taken the place of supernatural intervention; but in many of these stories and novels the religious machinery is deliberately exposed at the

---

[10] Translations: Arthur Waley, *Monkey* (abridged) (New York, 1943); Timothy Richards, *A Mission to Heaven* (partial, with summaries) (Shanghai, 1913).

[11] For examples, see *Ku-chin hsiao-shuo*, 1, 3, 38; and *Hsing-shih heng-yen*, 15.

end. When a character declares that a certain event is the result of karma, we are prepared to accept this as a social convention debased almost to a figure of speech and do not allow it to influence our understanding of the plot. When, however, the narrator himself concludes so excellent a story as "The Ring" in *Ch'ing p'ing shan t'ang hua-pen* with an explanation that the whole of the tragic sequence we have just accepted imaginatively as true, is really the result of bad karma stemming from previous incarnations of the two main characters; when he says in effect that the development of the tragic situation has all the time been wholly outside the power of the hero and heroine to control or alter, the Western reader feels that he has been imposed upon and tricked. In the same way he objects to being harrowed by the troubles of Tess of the D'Urbervilles while being told all along that they are merely a cruel jest on the part of the Immortals. Granted that karma is an alien belief in the West, I suspect that even the Chinese reader who may have accepted the explanation as religiously sound, will nevertheless, feel an aesthetic disappointment at its unnecessary use in such a literary context.

If the principle of ultimate motivation in Chinese fiction is ambiguous, its moral purpose is equally so. Much of its traditional narrative material is frankly pornographic or immoral in nature, and much of what remains is amoral inasmuch as it unconsciously pictures a world governed not by the moral order of philosopher or priest, but by the operation of blind chance. The onus of immorality by which fiction was traditionally regarded in China as detrimental to the morals of society, is, therefore, not entirely without justification. To circumvent criticism by Confucian officialdom, writers stressed the value of fiction as moral instruction and missed no opportunity to include homilies on the Confucian virtues and thus provide a specious pedagogic function which is wholly foreign to such literature. The result is a marked contradiction between the avowed and the implicit moral purpose which destroys that integrity we expect of good fiction.

In the matter of character protrayal, another contrast between Chinese and Western fiction is apparent. Both literatures attempt realistic portrayals of social types and the difference between them is one of degree. Both exploit dialogue as a means of differentiating character and caste. The novel of the West, however, explores more thoroughly the minds of characters, and long familiarity with this realm has made possible whole novels which are confined to the individual mind alone, such as those of Virginia Woolf and James Joyce. But to the Chinese novelist, the mental life of his fictional characters is an area to be entered only briefly when necessary and then with timidity. For this reason, his ability to exploit one of the chief concerns of realistic fiction, the discrepancy between appearance and reality, is severely limited since he can rarely show us the sharp variance between what is said and what is thought.

The writer of colloquial fiction, in spite of his keen eye for movement and his sharp ear for the speech of daily life, is curiously deficient in description of manner which requires a constant and subjective identification with one's imaginary characters. Translators soon notice this lack in the invariable use of *tao*[c] or *yüeh*[d] to introduce all speeches. This non-committal "said" often leaves the ensuing speech utterly colorless or ambiguous in tone. When in English such a statement as "I am going" is given emotional color by the verb which follows; as "'I am going,' she grumbled;" "'I am going,' she insinuated;" "'I am going,' she sighed; " or "'I am going,' she screamed;" the Chinese narrator gives no hint of the emotion implicit in each speech beyond that which the speech itself suggests. As a result, even the racy, supple and vital dialogue, which is one of the strong points of colloquial fiction, sometimes has the quality of monotone to the reader conditioned to the subtle overtones suggested by the stage directions in Western fiction.

This limitation of psychological analysis—which is what in general terms it is—seems to be related to a social factor influential in the development of Chinese fiction, and that is the absence of an aristocratic-feminine tradition in this branch of literature. I think most historians of French literature would agree that the psychological perceptiveness of the French novel can be traced to the influence of Mme. de La Fayette and Mlle. de Scudéry as well as their many contemporaries. Certainly the bulk of English fiction from the seventeenth century into the nineteenth was the work of women—not those novels which are read today, but those which were widely admired at the time and which prepared the way for the work of Richardardson and his successors. Characteristic of this feminine and quite often aristocratic tradition is a preoccupation with minute analysis of emotions and probing of mental attitudes. In a wholly unrelated literature, one has only to think of the *Genji monogatari*[12] by Lady Murasaki for a strikingly similar example of psychological sensitivity in the work of a woman and one associated with a sheltered and highly sophisticated court circle.

Whether the Chinese civil service system, restricting the growth of a permanent aristocracy isolated from the world of action, or whether the absence of women writers in either the literary language or colloquial genres of fiction, are factors which can account for the psychological immaturity of characterization in novels, is a thesis I am not prepared to defend. As a theme for speculation and study, however, it would be of value in apparaising the character of the Chinese novel. Any such speculation must take into account the one isolated specimen of psychological sophistication, *Hung lou meng*,[13] and determine its relationship to the main stream of Chinese fiction.

From the preceding remarks it should be evident that the genesis of a realistic fiction in China and in Europe had many features in common: appeal to a lower class audience uninterested in a past classical tradition; material which in revolt to that tradition was earthy and sensational instead of intellectual and restrained; and a narrative style that was sensuous rather than symbolic, observant rather than contemplative. Yet in spite of the similar origins, Western fiction, freeing itself early from the odium of immorality and confining itself to

---

[12] Translation: Arthur Waley, *The Tale of Genji* (Boston, 1927–33).

[13] Translations: H. Bancroft Joly, *Hung Lou Meng or, The Dream of the Red Chamber, a Chinese Novel* (partial) (London, 1892–93); Chi-chen Wang, *Dream of the Red Chamber* (partial, with summary) (New York, 1912).

the realm of mundane fact, was able to progress further in the chosen direction and explore possible byroads. Chinese fiction, on the other hand, constantly defending its right to exist, hampered by anachronistic materials and stylistic conventions, and unable to face frankly the direction in which it tended, traveled more slowly and fitfully along the same road toward realism until the influence of Western models began to be felt at the end of the nineteenth century.

In concluding, I hope my main intention has been clear. Despite the inevitability of a value judgment, that intention has not been to disparage a great tradition of fiction in China, but rather to further in the general reader an appreciation of its works in translation by suggesting what he must not expect of it. Understanding and accepting its unfamiliar conventions, he will find in its works much profit, diversion and an admirable craftsmanship in the art of story-telling.

a 回	*k* 東京夢華錄	*t* 警世通言
b 今古奇觀	*l* 唐宋叢書	*u* 清平山堂話本
c 道	*m* 耐得翁	*v* 洪便
d 曰	*n* 都城紀勝	*w* 京本通俗小說
e 段成式	*o* 楝亭十二種	*x* 古今小說
f 西陽雜俎續集	*p* 周密	*y* 喻世明言
g 湖北先正遺書	*q* 武林舊事	*z* 醒世恒言
h 蘇軾	*r* 筆記小說大觀	*aa* 馮夢龍
i 東坡志林	*s* 詞	*ab* 三言
j 孟元老		

ADDITIONS AND CORRECTIONS

Page 246, note 13, line 3: *For* (New York, 1912) *read* (New York, 1958) *and add* Franz Kuhn, *Dream of the Red Chamber* (tr. by Florence and Isabel McHugh, New York, 1958).